AF333985

Questions & Answers in QUARTZ WATCH REPAIRING

by

Alice B. Carpenter, CMW, CEWS

and

Buddy Carpenter, CMC, CEWS

American Watchmakers Institute Press

AWI

Copyright ©1989 by the American Watchmakers Institute

All rights reserved. No part of this book may be reproduced or transmitted in any form or by any means, electronic or mechanical, including photo-copying, recording, or by any information storage and retrieval system, without permission in writing from the publisher.

Printed in the United States of America
Library of Congress Catalog Card Number 89-082038
ISBN Number 0-918845-13-0

Contents

SECTION I

Terms, Definitions & Symbols

Alternating Current: an electrical current that changes its polarity at fixed intervals. Abbreviation or symbol: A.C.

Ampere: a unit of electric current; the amount of current passing through a circuit with a resistance of one ohm with an electromotive force of one volt. Abbreviation or symbol: A

Analog: in timepieces, the use of hands to display the time.

Anode: the negative terminal of a cell. Abbreviation or symbol: −

Base: the center section of a transistor. The N part of a PNP, and the P part of an NPN transistor. See *transistor*.

Battery: a series of cells linked together, either in series or parallel. Abbreviation or symbol:

Battery Clamp: the metal strap or clamp securing the battery to a movement.

Capacitor: a temporary reservoir of electrons. See *condenser*. Abbreviation or symbol:

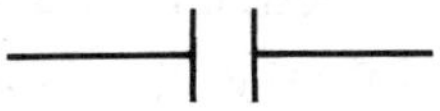

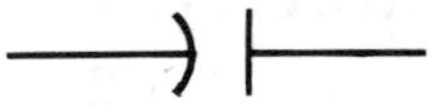

Cathode: the positive terminal of a cell. Abbreviation or symbol: +

Cell: a device, primarily electrochemical, for converting other kinds of energy into electrical energy, usually in the range between 1 and 1½ volts. A primary cell. Abbreviation or symbol:

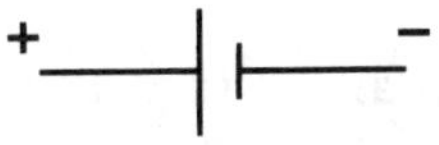

Charge: usually refers to the accumulation of electrons on the plates of a capacitor or electrodes of a battery.

Coils: a spool of insulated wire wound on a core through which an electric current is passed to create a magnetic field. Abbreviation or symbol:

Collector: that part of the transistor that collects the current carriers which come from the emitter of that transistor.

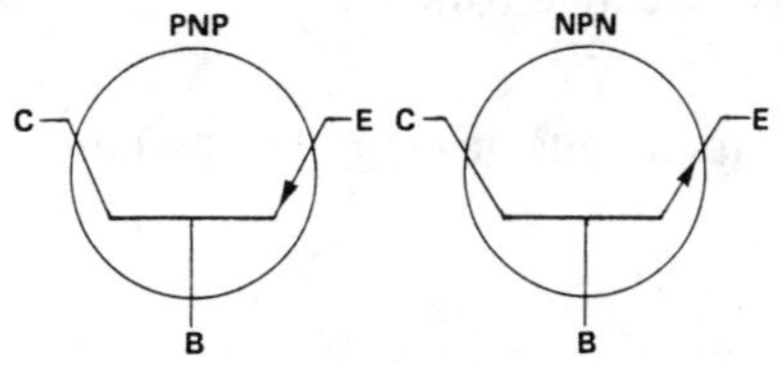

Condenser: a device for accumulating an electrical charge. See *capacitor*.

Conductor: in electricity, a material which allows an electric current, thus electrons, to pass through it freely.

Conductance: the degree of efficiency of a substance to allow a current to flow.

Conductive Epoxy: a two-part adhesive, one part of which contains sufficient conductive particles to form an electrical path after the adhesive has set and hardened. Used in the repair of cut, broken, or interrupted open circuits.

Consumption: in electric watches and clocks, the amount of power drawn from the battery or energy source to keep the timepiece operating.

Crystal: see *quartz crystal*.

Current: movement, flow, propagation of electrons in a certain direction. The unit in electricity measured in Amperes.

Cycles: series of phenomena or functions that are repeated.

Diode: an electronic valve which permits electricial current to flow through it but in one direction. Abbreviation or symbol:

Divider Circuit: a section of the integrated circuit which divides the output frequency of the quartz crystal, generally by 2, at each divider step downward, to enable an output pulse of one per second, per minute, per hour, day, month, etc.

Direct Current: produced electrochemically by a cell, its electrons do not flow in cycles as alternating current does. Abbreviation or symbol: DC

Dry Cell: a cell that has a metal covering, a metal cap, and a seal, such as the cells that are used in watches.

Eddy Currents: a current moving in a contrary direction to the main current.

Electrical Shunt: a bridge made of certain resistive metal alloys to collect and redirect current flow or bypass its normal route.

Electrode: a pole or terminal on electrical equipment, such as in an energy cell, capacitor, or diode.

Electrolyte: refers to chemical solutions used in batteries and sometimes in special kinds of capacitors called electrolytic capacitors.

Electromagnet: a magnet that loses its magnetic properties when the application of voltage ceases.

Electromotive Force: the force producing an electric current. Abbreviation or symbol: EMF

Electron: the negatively charged particles of an atom.

Emitter: the part of the transistor which emits the current carriers. One of the two NN or PP leads in a transistor, the other being the collector.

Farad: the unit of electrical capacity of a capacitor. A farad is a charge equal to one coulomb at one volt. Symbol: F

Ferrous: of or containing iron; can be easily magnetized.

Flip-Flop: like a teeter-totter, an electronic combination of two circuits with the same characteristics and prepared in such a manner so that when one is in the active mode the other is blocked. Used in divider circuits.

Frequency: the number of oscillations per unit of time.

Frequency Divider: an integrated circuit or computer which divides oscillations or frequency down to a lower frequency.

Fuse: Abbreviation or symbols:

Ground: Abbreviations or symbols:

Hertz: oscillations or cycles per second.

Induced Voltage: when a magnetic field is caused to move over a coil or wire, electrical voltage and currents are created; these are said to be induced.

Insulator: a non-conductor, used to prevent the passage of current.

Integrated Circuit: an electric circuit planned on the drawing board and reduced microscopically by photographic process deposited on a silicon crystal semiconductive surface, creating with sequential applications to become transistors, resistors, capacitors with their connective paths, etc. Abbreviation or symbol: IC

Infinity: the figure on the extreme left of the ohms scale would be an infinite resistance. The opposite of a short circuit reading. Abbreviation or symbol: ∞

Junction: the area where semiconductor materials are joined together.

Law of the Magnets: opposite poles attract, like poles repel.

Light-Emitting Diode: a diode composed of gallium arsenide and phosphide that glows red when a voltage is applied. The readout used most often on early digital quartz watches. Abbreviation or symbol: LED

Liquid Crystal Display: a display panel containing a liquid having the optical properties of a crystal between 0 degrees and 70 degrees C. The liquid changes its molecular order when under electrical stimulus to effect a digital display. Abbreviation or symbol: LCD

Magnetic Shunt: a plate or bridge whose purpose is to collect and redirect the magnetic field; some are used to protect the movement from outside magnetic influence that might damage the circuitry, while others are used to protect some part of the movement from inside magnetic influence, such as the shunt bridge in an electric Timex balance wheel movement, where it protects the hairspring from the influence of the magnets and coil in the watch.

Magnetic Field: the area which the magnetic lines of force cover.

Magnetic Lines of Force: invisible, definite pattern emanating from any magnet.

Meter: Abbreviation or symbol:

Microampere: the millionth part of an Ampere. Abbreviation or symbol: μA

Milliampere: the thousandth part of an Ampere. Abbreviation or symbol: mA

Milliampere Hour: the potential energy measured in thousandths of an Ampere passing in one hour. Abbreviation or symbol: mAh

Multimeter: a combination meter which reads volts, amps, and ohms.

Negative Terminal: the anode of a cell. Abbreviation or symbol: −

Nonferrous: containing no iron; therefore, not magnetizable.

Ohm: a unit of electric resistance. Abbreviation or symbol: Ω

Ohm's Law: the basic relationship between volts, amps, and ohms.

$$\frac{E}{I \mid R}$$

Oscillation: the definite frequency of the forward and back rushing of electric charges in a conductor or circuit possessing both inductance and capacitance.

Oscillator: that part of the electronic circuit resulting in high, fixed frequency alternating currents. Quartz crystals fall into this category.

Pegging: a term used to describe the action of the needle of the meter when it strikes the stop at the extreme range.

Piezo Electric: the property of a substance to vibrate when under the influence of an electrical current and emit an alternating current when under mechanical pressure. Quartz crystal and certain ceramic transducers as in alarm watches have these properties.

Positive Terminal: the electrode to which the electron stream flows, as the cathode of a cell. Abbreviation or symbol: **+**

Printed Circuit: a circuit for electronic apparatus made by depositing conductive material in continuous paths from terminal to terminal on an insulting surface.

Permeability: the ability of a metal to gather in the magnetic lines of force.

Permanent Magnet: a magnet that maintains its magnetic properties almost indefinitely.

Polarity: one end of the magnet having a south pole, or negative magnetic field, and the other end of the magnet having a north pole or a positive magnetic field.

Pulse: a signal or series of short duration.

Quiescent Current: the small current flowing through an analog or digital quartz watch counting time between step motor impulses.

Quartz Crystal: a mineral found in nature, also produced synthetically. A frequency standard possessing piezo electric properties. Electronic device equivalent to balance assembly in mechanical watch.

Resistance: the characteristic of a material to retard, restrict, or prevent the flow of an electric current. Usually a purposely placed resistance to current flow.

Resistor: any material, device, that offers resistance to the flow of electric current. Abbreviation or symbol: ⎓WWW⎓

Retrofit: a unit, module, movement made to take the place of an original.

Semiconductor: certain materials, namely silicon and germanium, can be combined with special impurities and they exhibit special conducting or insulating abilities depending on the polarities to which they are connected.

Short Circuit: usually an unintentional condition of low resistance between two points of different polarity in a circuit—results in a flow of excess current.

Stepping Motor Quartz: Abbreviation: SMQ

Solid State: having no moving parts.

Transistor: composed of three chambers, in watches it is used as a switching device. It allows either negative or positive current to flow in one direction, depending on the kind of transistor it is. The two kinds of transistors are (abbreviations or symbols):

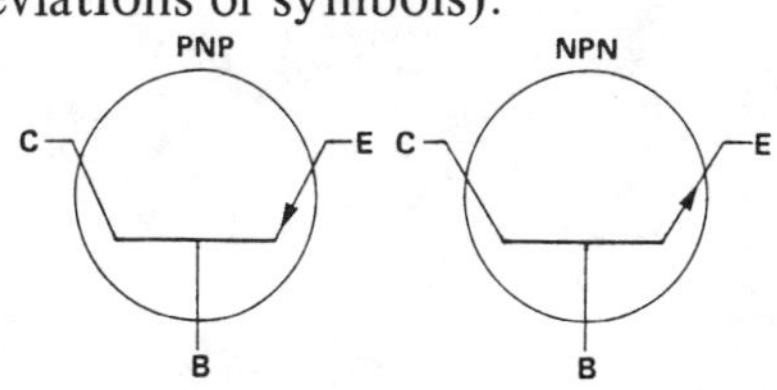

Trimmer Capacitor: used in quartz watches to alter the frequency or regulate the timekeeping. Abbreviation or symbol:

Thousands: on the meter the letter K is used to denote thousands.

Temporary Magnet: maintains its magnetic properties only while under the influence of another magnet.

Volt: a unit for measuring the force exerted to produce an electric current. Abbreviation or symbol: V (on the meter, we read AC volts, or DC volts).

Voltage: electrical force measured in volts.

VOM: the letters stand for Volt, Ohm, Milliamp, Meter. See *multimeter*.

Wet Cell: a cell that water can be added to, such as the cell or battery used in cars.

Wires Connected: abbreviation or symbol:

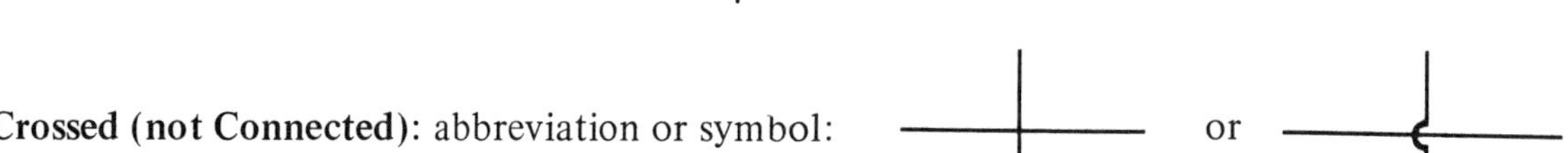

Wires Crossed (not Connected): abbreviation or symbol: or

Zebra Connectors: elastomer bar with alternate black and white stripes used as segment tab-connectors in an LCD. Used to connect the circuit to the display.

SECTION II

Multiple Choice Questions

_____ 1. If the current flow is reversed in an electromagnet:

    a. the magnetic fields are reversed
    b. negative current flow is blocked
    c. positive current flow is blocked

_____ 2. The variable capacitor connected to the quartz crystal in order to regulate it is:

    a. a capacitor
    b. a condensor
    c. a trimmer capacitor

_____ 3. In a circuit, a diode will allow:

    a. current to flow in either direction
    b. current flow in one direction only
    c. current to flow from positive to negative

_____ 4. The definite, invisible pattern emanating from an electromagnet or permanent magnet is:

    a. magnetic lines of force
    b. the law of magnets
    c. permeability

_____ 5. The anode of a cell is:

    a. the positive terminal
    b. the negative terminal
    c. neither of the above

_____ 6. The cathode of a cell is:

    a. the positive terminal
    b. the negative terminal
    c. neither of the above

_____ 7. The resistance of a wire is governed by:

    a. movement, flow, propagation of current
    b. its length, thickness, composition, and temperature
    c. the ability to gather in magnetic lines of force

_____ 8. A battery is:

    a. a single cell
    b. two or more cells
    c. cells connected in parallel

_____ 9. The purpose of the booster coil is:

    a.   to increase voltage to the alarm circuit
    b.   to increase current to the alarm circuit
    c.   to provide power to the stepping motor

_____ 10. The purpose of a capacitor is:

    a.   it is a temporary storage of electrons
    b.   it will allow current to flow only in one direction
    c.   it is an electronic switch

_____ 11. An ampere is:

    a.   a unit of measure of current flow
    b.   a unit of measurement of resistance
    c.   a unit of measurement of capacitance

_____ 12. Ferrous means:

    a.   of or containing iron
    b.   containing no iron
    c.   neither of the above

_____ 13. Nonferrous means:

    a.   of or containing iron
    b.   containing no iron
    c.   neither of the above

_____ 14. Base, emitter, collector are:

    a.   parts of a transistor
    b.   parts of a conductor
    c.   parts of a diode

_____ 15. An electrical component that allows current to flow but in one direction only:

    a.   diode
    b.   condenser
    c.   coil

_____ 16. Sparking is caused by:

    a.   a backwash of negative current
    b.   an excessive flow of positive current
    c.   alternating current

_____ 17. An insulator allows:

    a.   current to flow freely
    b.   no current to flow
    c.   measured amounts of current to flow

_____ 18. A conductor allows:

    a.   current to flow freely
    b.   little or no current to flow
    c.   measured amounts of current to flow

_____ 19. Solid state means:

    a.   made of solid metal
    b.   having no moving parts
    c.   having mechanical switches

_____ 20. A unit for measuring the force exerted to produce an electric current is:

    a.   amp
    b.   volt
    c.   multimeter

_____ 21. If an analog quartz watch was brought in for a power cell and the old one is missing, which cell should be used if the watch has no additional functions?

    a.   high drain cell
    b.   low drain cell
    c.   mercury cell

_____ 22. If a quartz watch were brought in for a power cell and the old one is missing, and it has an alarm, would you use:

    a.   a high drain cell
    b.   a low drain cell
    c.   a mercury cell

_____ 23. Should a trimmer capacitor become broken loose from the circuit, the crystal frequency would:

    a.   always decrease
    b.   remain the same
    c.   increase by a small amount

_____ 24. Which of the following would indicate the trimmer has become disconnected or broken?

    a.   would change the crystal to a slower frequency
    b.   would make no change in frequency when the trimmer capacitor is turned
    c.   watch would show a fast rate of several seconds a day

_____ 25. Most of the current used by a step motor quartz watch is consumed by the:

    a.   oscillating circuit
    b.   dividing circuit
    c.   step motor

_____ 26. The voltage of the mercury cell is:

    a.   1.35 V
    b.   1.5 V
    c.   1.55 V

_____ 27. If a train wheel is missing, the timekeeping element of an electric/electronic watch will:

    a.   gain time
    b.   lose time
    c.   remain the same

_____ 28. When electrical current flows in a material that has little conductivity:

    a.   greater amperage is required
    b.   little voltage is required
    c.   great resistance is present

_____ 29. Two wire conductors made of the same material, and equal in size, length, and temperature would:

    a.   have like resistance
    b.   have different resistance
    c.   have slightly different resistance

_____ 30. When current electricity flows through a wire conductor:

    a.   a magnetic field is sometimes created in the wire
    b.   a magnetic field is always created in the wire
    c.   a magnetic field is created in the wire only if A.C. is used

_____ 31. Current is:

    a.   the flow of protons through a conductor
    b.   the flow of atoms through a conductor
    c.   the flow of electrons through a conductor

_____ 32. A conductor is any material which:

    a.   resists the flow of electrons
    b.   tends to become hot when electrons flow freely
    c.   allows electrons to flow easily

_____ 33. When increased voltage is applied to a conductor, the result will be:

    a.   reduced current flow
    b.   greater current flow
    c.   less resistance

_____ 34. D.C. current flows:

    a.   in one direction and then the other
    b.   in one direction
    c.   from positive to negative

_____ 35. A.C. current flows:

    a.   in one direction and then the other
    b.   in one direction
    c.   from positive to negative

_____ 36. When voltage is applied to a coil of insulated wire, the magnetic field created would:

    a. not change polarity regardless of the direction of current flow
    b. have no polarity
    c. reverse its polarity if current flow is reversed

_____ 37. Since all conductors have resistance, some of the energy is lost through:

    a. absorption
    b. neutralization
    c. heat

_____ 38. In a coil where the insulation is damaged allowing many insulated wires to touch, the result would be:

    a. less resistance, therefore a greater magnetic field
    b. greater resistance, therefore a greater field
    c. less resistance, therefore a smaller magnetic field

_____ 39. An electromagnetic field:

    a. is always present in an electromagnet
    b. is created only when applying current
    c. requires no applied voltage

_____ 40. Temporary magnets:

    a. are never used as a core in an electromagnet
    b. are usually made of soft iron or similar material
    c. are made of carbon steel

_____ 41. Resistance is:

    a. the amount of electrical pressure
    b. the rate of flow of electrons
    c. the amount of opposition to the flow of electrons

_____ 42. An energy cell can be tested with a:

    a. voltmeter
    b. ammeter
    c. ohmmeter

_____ 43. The basic function of a resistor is to:

    a. act as a one-way valve
    b. store electrons which can be released suddenly
    c. reduce current flow, but allow a predetermined amount to pass

_____ 44. A resistor is sometimes used to:

    a. create a bias condition in the base of a transistor
    b. to hold a charge somewhat like a battery
    c. to step up voltage

_____ 45. When the negative terminal of a battery is connected to the positive side of a diode, and the positive terminal of a battery is connected to the negative side:

    a.   current will flow
    b.   current will not flow
    c.   current will be alternating

_____ 46. The capacitor:

    a.   is used as an electronic switch
    b.   reverses the flow of current
    c.   stores energy somewhat like a battery

_____ 47. A capacitor:

    a.   prevents current flow in one direction
    b.   has the ability to give off a greater, more sudden surge of current than the energy cell used to create its charge
    c.   is used in conjunction with mechanical switches

_____ 48. The major problem with the electric watch was:

    a.   it used excessive current
    b.   it was too complicated for the watchmaker to repair
    c.   the mechanical contacts which served as switching devices burned and pitted rather quickly causing malfunction

_____ 49. A transistor with a negative base is called:

    a.   NPN type
    b.   PNP type
    c.   PNNP type

_____ 50. When any electric/electronic watch is brought in for repair, the first check the watchmaker should make is:

    a.   the coils
    b.   the cell
    c.   the circuit

_____ 51. To simplify troubleshooting electronic analog watches, the watchmaker should:

    a.   determine if watch is dirty
    b.   check the coils
    c.   separate the mechanical from the electronic possibilities

_____ 52. The electronic reduction of the crystal frequency 32,768 to one pulse per second is accomplished by:

    a.   using a transistor switch
    b.   applying battery voltage to the crystal
    c.   a series of 15 flip-flop circuits

_____ 53. When there is no input voltage on the base of a transistor:

     a.   it takes longer to complete its function
     b.   it won't work
     c.   it has excessively high collector current

_____ 54. A piezoelectric effect occurs:

     a.   when a hexagon-shaped crystal is cut in a very thin slice
     b.   when a crystal develops electrical voltages across opposite faces
     c.   when voltage is applied to the quartz crystal

_____ 55. The advantage of the LCD watch over the LED was:

     a.   brighter display
     b.   more accurate
     c.   display on continuously

_____ 56. A "zero" reading on the ohmmeter means:

     a.   a half-dead cell
     b.   a shorted coil
     c.   an open circuit

_____ 57. In a quartz watch that has a regulator, it is:

     a.   a resistor which controls battery voltage to the crystal
     b.   a spring pressing against the crown
     c.   a capacitor connected to the crystal

_____ 58. An insulator would be used to:

     a.   speed up the flow of electrons
     b.   increase the voltage
     c.   block the flow of electrons

_____ 59. The capacitor:

     a.   is called an electronic switch
     b.   has the ability to give off a greater more sudden surge than the power source
     c.   is only used as a trimmer capacitor

_____ 60. A resistor:

     a.   acts as a valve
     b.   stores electrons which can be released suddenly
     c.   is used to reduce current flow, but allow a predetermined amount to pass

_____ 61. In a stepping motor quartz watch, the stator is:

     a.   a permanent magnet
     b.   is not a magnet
     c.   part of an electromagnet

_____ 62. Voltage is:

    a.   the rate of flow of electrons
    b.   the amount of resistance to the rate of flow of electrons
    c.   the amount of electrical pressure

_____ 63. An electromagnet which consists of just a coil of insulated wire:

    a.   can be permanently magnetized
    b.   will develop a weaker field if more current is applied
    c.   reverses polarity when current is reversed

_____ 64. The influence that a permanent has on an electromagnet is:

    a.   increased as they are spread farther apart
    b.   decreased as they are spread farther apart
    c.   decreased as they are brought closer together

_____ 65. The type of magnet that has no magnetism until electric current is applied is:

    a.   permanent magnet
    b.   electromagnet
    c.   nonmagnetic

_____ 66. When a magnet is moved in one direction near a coil of wire that the magnetic field cuts through the wires:

    a.   the magnet gets weaker
    b.   the polarity is reversed
    c.   current flows in the coil

_____ 67. If a magnet is held stationary and a coil of wire is moved near it:

    a.   the coil will get warm if it is moved too fast
    b.   the magnet gets weaker
    c.   the magnetic field will reverse if the motion of the coil reverses

_____ 68. The purpose of an iron core in an electromagnet is:

    a.   to hold the coils of the wire more securely
    b.   to increase the current
    c.   to concentrate the magnetic field

_____ 69. The main reason the quartz timepieces are more accurate than balance wheel types is:

    a.   quartz is very hard and requires no oiling
    b.   quartz crystals can operate at very much higher rates than balance wheels
    c.   quartz cut on an axis are not temperature sensitive

_____ 70. To check the coil that is a part of the circuit board you should remove the cell from the watch, place your meter probes on the coil connection pads, and:

    a.   set your meter function switch to RX100 and read the resistance on the coil
    b.   Supply .2 volts from your variable power supply and with your meter function switch set to the micro amps scale, read current flow
    c.   with your meter function switch set on your lowest D.C. voltage scale, read voltage

_____ 71. To check the output of the circuit of a stepping motor watch you would set your meter function switch to the lowest D.C. voltage scale and place the meter probes:

    a. on the positive and negative cell contacts
    b. on the coil connection pads
    c. on the quartz crystal connections

_____ 72. If a low drain cell were placed in a watch containing a night light and an alarm:

    a. the watch would not run
    b. the watch would run, but the alarm would not operate
    c. the watch would run, but would lose time

_____ 73. A stepping motor watch should be expected to run with the voltage reduced to:

    a. 50% of normal cell voltage
    b. 80% of normal cell voltage
    c. 90% of normal cell voltage

_____ 74. When checking a stepping motor for current consumption, if a capacitor were placed in parallel to the power supply, at each impulse, the meter needle would move:

    a. farther upscale
    b. less upscale
    c. have no effect at all

_____ 75. When checking the resistance of a coil that has been separated from the circuit board you find that it reads less than 500 Ohms, you should:

    a. replace the coil
    b. attempt to repair the coil
    c. suspect that the circuit board is defected

_____ 76. When checking the coil for resistance you get no meter deflection, you would:

    a. reverse the meter probes
    b. consider the coil to be shorted
    c. consider the coil to be open

_____ 77. In checking the output of the circuit board you would see the meter needle:

    a. move up to 1.55 volts and remain there
    b. move upscale and downscale alternately in sync with the energizing of the coil
    c. move upscale with each pulse of current to the coil

_____ 78. If you placed a high drain cell in a watch which required only a low drain cell:

    a. it would ruin the watch
    b. it would cause the watch to gain time
    c. it would cause no malfunction of the watch, but the cell might not last quite as long

_____ 79. A conductor is any material which:

    a.   greatly resists the flow of electrons
    b.   allows electrons to flow freely
    c.   tends to overheat when any electrons flow in it

_____ 80. The temperature stability of a quartz crystal for timekeeping purposes depends on:

    a.   its period of time spent in the aging process
    b.   its location in the circuit
    c.   the battery voltage of the watch

SECTION III

True-False

_____ 1. An electromagnet is a temporary magnet.

_____ 2. The cell invented by Allesandro Volta was a wet cell.

_____ 3. Most 1.55 volt watch cells are wet cells.

_____ 4. The term "battery" originally meant one cell.

_____ 5. The core of an electromagnet is made of steel.

_____ 6. Like magnetic poles attract and unlike poles repel.

_____ 7. Not all magnets have polarity.

_____ 8. The quartz crystal in an electronic watch serves the same function as the mainspring in a mechanical watch.

_____ 9. The ohmmeter is used to measure current.

_____ 10. The ammeter is used to measure current.

_____ 11. A short circuit is usually an intentional conduction of low resistance between two points of different polarity in a circuit.

_____ 12. The terms "capacitor" and "condenser" mean the same thing.

_____ 13. The divider circuit divides the output of the quartz crystal to a usable frequency.

_____ 14. An electromagnet uses a soft iron core to increase the length of the battery life.

_____ 15. Coils are composed of insulated wire.

_____ 16. The basic function of a resistor is to reduce current flow but allows a predetermined amount of current to pass through the circuit.

_____ 17. Electromagnets are composed of coils of insulated wire, with a core of soft iron.

_____ 18. Distilled water is a good conductor.

_____ 19. When reading current consumption, the meter is set on the resistance scale.

_____ 20. If cells are connected in series, the voltage remains the same.

_____ 21. Paper is a good nonconductor.

_____ 22. In order of superiority, silver is the best conductor.

_____ 23. Ferrous metals can be easily magnetized.

_____ 24. Nonferrous metals can be easily magnetized.

_____ 25. A watch cell is a dry cell.

_____ 26. The metal in the core of an electromagnet is of temporary magnetic material.

_____ 27. If cells are connected in parallel, the voltage remains the same.

_____ 28. The resistance of a wire is governed by the amount of current applied.

_____ 29. A diode is composed of a base, emitter, and collector.

_____ 30. If a quartz watch is brought in for a cell, and the cell is missing, the fact that the watch had an alarm would indicate a high drain cell is needed.

_____ 31. If a quartz watch is brought in for a cell, and the cell is missing, the fact that it is a stepping motor with no functions (such as s.s. hand, alarm, light, etc.) would indicate that a high drain cell is needed.

_____ 32. Magnetic fields decrease in strength with distance from their source.

_____ 33. Hitting a magnet could demagnetize it.

_____ 34. A transistor is a switch.

_____ 35. The timekeeping element of an electric or electronic watch will oscillate even if the center wheel is missing.

_____ 36. Moving a conductor across a magnetic field produces a voltage in the conductor.

_____ 37. The voltage applied to the base of a transistor is called "bias" voltage.

_____ 38. In an electric watch, the cell takes the place of the mainspring.

_____ 39. The volt is a measure of electrical pressure.

_____ 40. Amperage is the unit of electric rate of current flow.

_____ 41. A conductor is any material which resists the flow of electrons.

_____ 42. When excessive voltage is applied to a wire conductor, the result could be melting of the wire.

_____ 43. When an insulated wire conductor is wound into a coil and voltage is applied, the magnetic field created would have a north and south polarity.

_____ 44. If you have a 1.35 volt cell and a 1.55 volt cell, the 1.55 volt cell would cause fewer electrons to flow in a given period if the resistance were the same.

_____ 45. If you have two 1.55 volt cells, but one is twice the size of the other, the smaller cell would produce less pressure or force.

_____ 46. Two cells, both the same voltage but different in size, would produce the same current flow for the same length of time.

_____ 47. A magnetic field always has at least one north and one south pole.

_____ 49. A magnetic field is made up of magnetic lines of force or flux.

_____ 50. A magnetic field can be collected and re-directed with certain kinds of metals.

_____ 51. Permanent magnets will retain their magnetism almost indefinitely unless demagnetized by a stronger magnet or demagnetizer.

_____ 52. All conductors have some resistance.

_____ 53. Temporary magnets do not have polarity when current is applied.

_____ 54. Temporary magnets are usually made of soft iron or similar material.

_____ 55. Electromagnets are magnetic only when current is flowing through them.

_____ 56. Electromagnets which consist of just a coil of wire can be permanently magnetized.

_____ 57. Electromagnets are made stronger by the addition of more coils of insulated wire.

_____ 58. An electromagnet has a north and a south pole when current is applied.

_____ 59. An electromagnet has no magnetic field when current is cut off.

_____ 60. An electromagnet reverses polarity when current is reversed.

_____ 61. An electromagnet will generate a minute amount of current when passed over, through, or near a permanent magnet.

_____ 62. An electromagnet field always travels in the same direction regardless of which way the current flows.

_____ 63. The basic function of a resistor is to act as a one-way valve.

_____ 64. The basic function of a resistor is to store electrons which can be released suddenly.

_____ 65. A resistor is sometimes used to hold a charge somewhat like a battery.

_____ 66. A resistor will allow current to travel in one direction but not the other.

_____ 67. The basic function of a diode is to allow current to travel in one direction but not the other.

_____ 68. A capacitor acts as a one-way valve.

_____ 69. A capacitor has the ability to control and stabilize current flow.

_____ 70. A capacitor has the ability to give off a greater more sudden surge of current than the energy cell used to create its charge.

_____ 71. A basic function of the capacitor is to store energy somewhat like a battery.

_____ 72. In horology, the accepted basic difference between an electric and an electronic watch is use of a transistor as a switching device.

_____ 73. The electric watch was abandoned by most watch manufacturers because the mechanical contacts which served as switching devices burned and pitted rather quickly causing malfunctioning.

____ 74. The electric watch always contained mechanical switches or contacts.

____ 75. The most common use of a transistor is as an electronic switch.

____ 76. When no voltage biases the base of the transistor, the transistor will not work.

____ 77. The LCD consumes more power than the LED because the display is on constantly.

____ 78. The multimeter can detect an open circuit.

____ 79. A "zero" reading can be either an open coil or an open circuit.

____ 80. Steel should be used to make a permanent magnet.

____ 81. An electromagnet can attract a permanent magnet.

____ 82. A directional compass will not work in a brass case because the brass case will shield it from magnetism.

____ 83. An electromagnet retains its magnetism while not in use.

____ 84. Polarity is common to all magnets.

____ 85. A compass needle is a temporary magnet.

____ 86. One should never put an electric watch in the demagnetizer.

____ 87. If a piece of old clock spring were magnetized, it would become a temporary magnet.

____ 88. When a balance assembly contains a permanent magnet and is pulsed with energy from an electromagnet, provided it is connected to a train of wheels, it could be used to keep time.

____ 89. Magnetic lines of force will pass through brass.

SECTION IV

Matching Questions

_____ 1. A short circuit would result in:

_____ 2. A resistor in a circuit will cause:

_____ 3. A nonconductor in a circuit will cause:

a. more current flow

b. less current flow

c. no current flow

d. a magnetic flow

e. current flow in a certain direction

f. alternating current

_____ 4. A short circuit would result in:

_____ 5. A resistor in a circuit will cause:

_____ 6. An insulator in a circuit will cause:

a. a higher resistance reading

b. a lower resistance reading

c. no current flow

d. positive current flow

e. negative current flow

f. maximum voltage reading

_____ 7. Conductor

_____ 8. Ampere

_____ 9. Shunt

a. free electron flow

b. unit of current flow

c. bypass of current flow

d. no current flow

e. minimal current flow

f. restricted current flow

_____ 10. Anode

_____ 11. Cathode

_____ 12. Diode

     a.   may be connected in

     b.   allows current to flow in both directions

     c.   acts as an electronic switch

     d.   allows current to flow in one direction only

     e.   positive terminal of a cell

     f.   negative terminal of a cell

_____ 13. Series

_____ 14. Parallel

_____ 15. Cell

     a.   two or more cells with anode connected to anode and cathode connected to cathode

     b.   two or more cells with anodes connected to cathodes

     c.   a single electrochemical unit

     d.   two or more cells connected in either series or parallel

---

_____ 16. Multimeter

_____ 17. Ohmmeter

_____ 18. Ammeter

     a.   measures resistance only

     b.   measures amperage only

     c.   two or more meters sharing a common dial

     d.   measures magnetism

     e.   measures AC current only

     f.   measures DC current only

_____ 19. Ohms

_____ 20. Ampere

_____ 21. Volt

     a.   a unit of measure of electrical resistance

     b.   a unit of measure of electric current

     c.   a unit for measuring the force exerted to produce an electric current

     d.   an electrical power, measured in watts

     e.   a unit for measuring oscillations per second

     f.   a unit for measuring cycles per second

_____ 22. Current

_____ 23. Voltage

_____ 24. Resistance

a. the flow of atoms through a conductor

b. the amount of current consumed

c. the flow of electrons through a conductor

d. the amount of current pressure

e. the amount of electrical pressure

f. the amount of opposition to the flow of electrons

_____ 25. Temporary magnet

_____ 26. Permanent magnet

_____ 27. Electromagnet

a. has no magnetic field when current is cut off

b. has a carbon steel core

c. has a constant north and south pole

d. is a magnet only when under the influence of another magnet

e. has a built-in resistor

f. uses a trimmer capacitor

_____ 28. Amperes

_____ 29. Current consumption

_____ 30. Resistance

a. creates a bias on the base of a transistor

b. a unit of measure of current flow

c. is read on the ammeter

d. will allow current to travel in one direction

e. is read on the ohmmeter

f. acts as a one-way valve

24

SECTION V

Completion Questions

1. A unit of electric resistance is known as______________________.

2. A unit of electric current is known as______________________.

3. A unit for measuring the force exerted to produce an electric current is known as______________.

4. An electrical pressure measured in volts is known as______________________.

5. Current is________________, ______________, and ________________in a certain direction.

6. A______________________is an electronic valve which permits electrical current to flow through it, but in one direction only.

7. A______________________is a temporary reservoir of electrons.

8. The frequency of oscillations per second is known as______________________.

9. An object made of or containing iron is known as______________________.

10. If two or more cells are connected in______________________, the voltage will remain the same.

11. If two or more cells are connected in______________________, the voltage will increase.

12. A______________________is used as an electronic switch.

13. ______________________ is known as electromotive force.

14. Three 1.5 volt cells connected in series will produce______________volts.

15. Three 1.5 volt cells connected in parallel will produce______________volts.

16. Like magnetic poles______________________and unlike magnetic poles______________________.

17. If the current flow is reversed in an electromagnet, the magnetic poles in the coil will ______________________.

18. The terminals of a cell are designated______________________and ______________________.

19. The most commonly used frequency of a quartz crystal is______________hertz.

20. The ammeter dial reads in______________________.

21. The ohmmeter dial reads in______________________.

22. The voltmeter dial reads in_________________________.

23. Sparking causes_____________________________buildup.

24. Household current is known as_________________________.

25. A diode is composed of_____________________chambers.

26. A_____________________ _____________________usually occurs when two or more turns of a coil lose their protective coating and allow current to pass between them.

27. A_____________________________ is a material that allows electrons to flow freely.

28. Electromagnets are composed of coils of_____________________wire.

29. Copper is a good_____________________of electricity.

30. Current electricity is the flow of electrons through a_____________________

31. When electrical current flows in a material having little conductivity, greater_____________________ is present.

32. When current flows through a wire conductor a_____________________is created around the wire.

33. _____________________ is the best conductor.

34. _____________________ will retain their magnetism almost indefinitely unless demagnetized by a stronger magnet or demagnetizer.

35. In a coil where the insulation is damaged, allowing many uninsulated wires to touch, the result would be_____________________.

36. An electromagnet reverses_____________________when current is reversed.

37. Amperage is the rate of flow of_____________________.

38. _____________________is the amount of electrical pressure.

39. Electrons will not flow through a_____________________.

40. The basic function of a_____________________is to reduce current flow but allow a predetermined amount to flow.

41. A_____________________stores energy somewhat like a battery.

42. To simplify troubleshooting electronic watches, the watchmaker should always separate the _____________________ from the electronic possibilities.

43. A transistor, which acts as an electronic switch, is composed of_____________________chambers.

44. The base of a_____________________ must be biased to allow it to function.

45. In a quartz crystal watch, the regulator is a _______________________ connected to the crystal.

46. Current is the flow of _________________ through a conductor.

47. A_________________ coil will read zero on the ohmmeter.

48. Excessive_________________ applied to a wire conductor could result in melting the wire.

49. The larger the_______________, the longer the life or stored energy.

50. _________________ will retain their magnetism almost indefinitely.

51. _________________ have polarity only when voltage is applied.

52. _________________ is read on the voltmeter.

53. _________________ is read on the ohmmeter.

54. _________________ is read on the ammeter.

55. An_______________ watch is one that uses a mechanical switch.

56. Mechanical switches became burned and pitted as a result of _______________.

57. The_______________should always be the first thing checked when any electronic watch is brought in for repair.

58. One advantage of the_______________watch is that the display stays on continuously.

59. One disadvantage of the_______________watch is that a button has to be depressed to actuate the display.

60. Electrons always flow in a direction from _______________ to _______________.

61. Opposition to the flow of current is called_______________.

62. The center section of a transistor is called the_______________.

63. The unit used to describe the capacitance (or size) of a capacitor is_______________.

SECTION VI

Short Answer Questions

1. Name three kinds of magnets:

    a.________________________

    b.________________________

    c.________________________

2. Name one thing on which the polarity of an electromagnet depends.

    ____

3. If two north poles are brought close together, what effect will they have on each other?

    ____

4. The strength of an electromagnet depends on the core material, the amount of current flowing, and one more factor. What is it?

    ____

5. Soft iron is used to make what kind of magnet? ________________________________

6. Define permanent magnet:____

    ____

7. Define temporary magnet: ____

    ____

8. Define electromagnet:____

    ____

9. Give the purpose of the metal core in an electromagnet.________________________

    ____

10. Most cells, regardless of size, are rated at what voltage? ____________________

    ____

11. Three 1½ volts cells are connected so that all the positive terminals are tied together and all the negative terminals are tied together. What is the voltage of the combination?________________

12. Cells are connected in series to increase the voltage or to make them last longer. (State which is true:) ____

13. Can a current flow without the presence of voltage?____________________________

14. When measuring voltage and the meter registers a negative reading, what should you do?

    ____

15. Which is larger—10mA, or 100μA?________________

16. .015 Amp is how many mA?________________

17. 200μA is how many mA?______________

18. A resistor marked 2.2K is how many ohms?________________

19. Which function of your multimeter (ohms, milliamps, or volts) has its own built-in battery for power? ____

20. If an ohmmeter cannot be "zeroed," what does it need? ___________________________________

21. When using the ohmmeter to measure the value of a resistor, does it matter if the probes are reversed? ____

22. Is it the stator or rotor that has the permanent magnets in a stepping motor watch?

____

SECTION VII

Meter Reading Questions

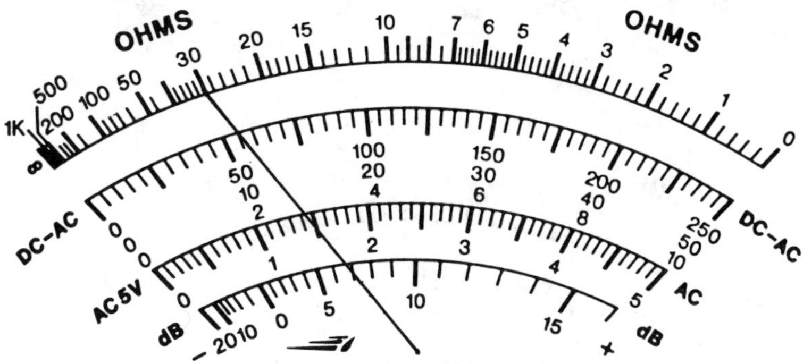

1. With meter set on RX 1 the correct reading is

_______________.

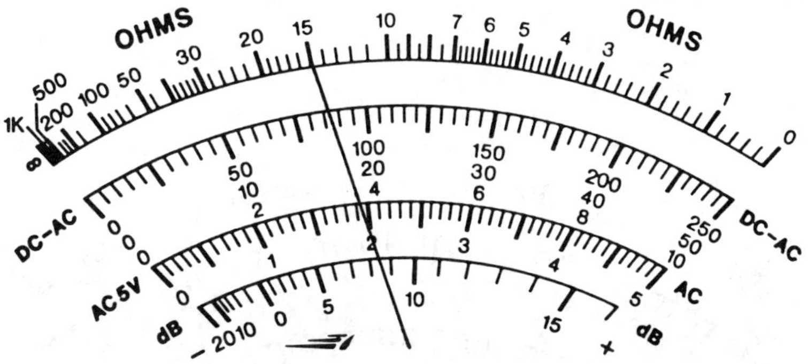

2. With meter set on RX 10 the correct reading is

_______________.

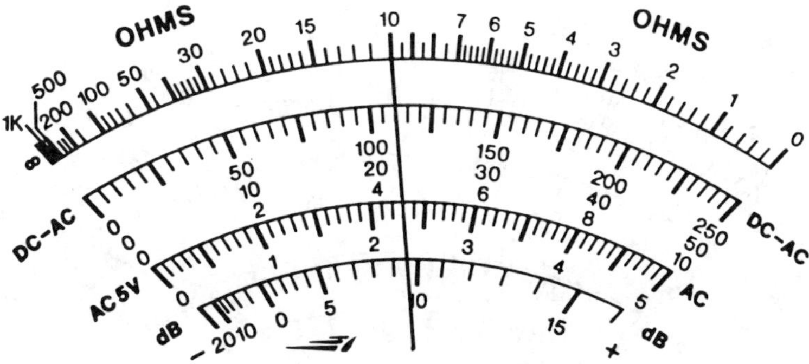

3. With meter set on RX 100 the correct reading is

_______________.

4. With meter set on RX 1K the correct reading is

_______________.

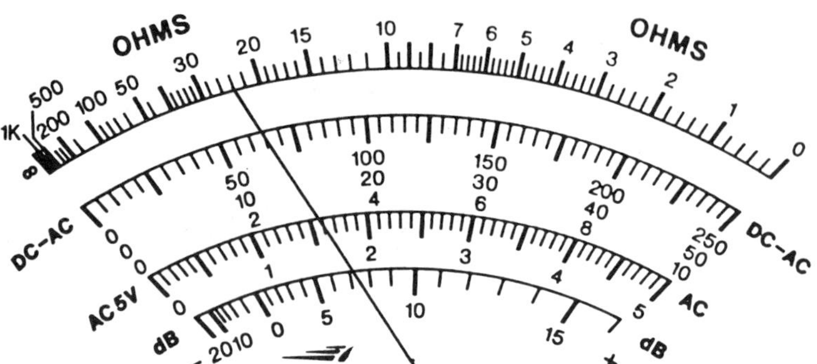

5. With meter set on 50 micro-amps the correct reading is

_______________.

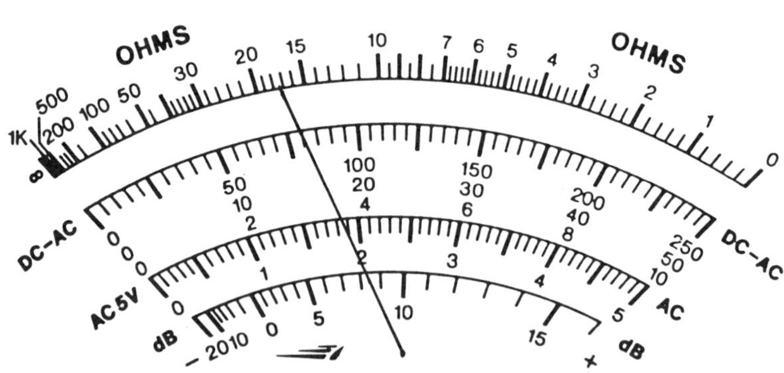

6. With meter set on 50 micro-amps the correct reading is

_______________.

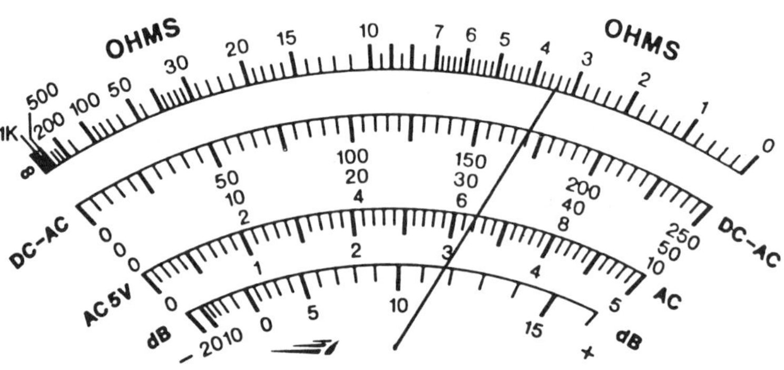

7. With meter set on 50 micro-amps the correct reading is

_______________.

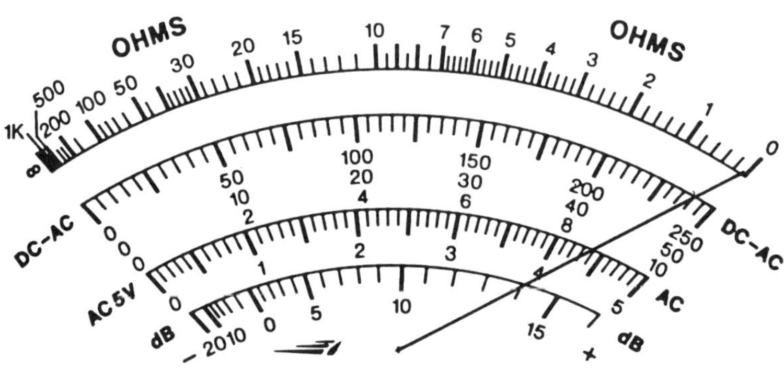

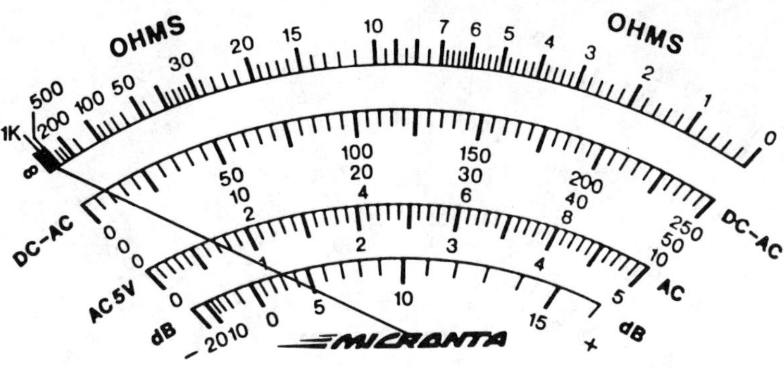

8. With meter set on 50 microamps the correct reading is

_______________________.

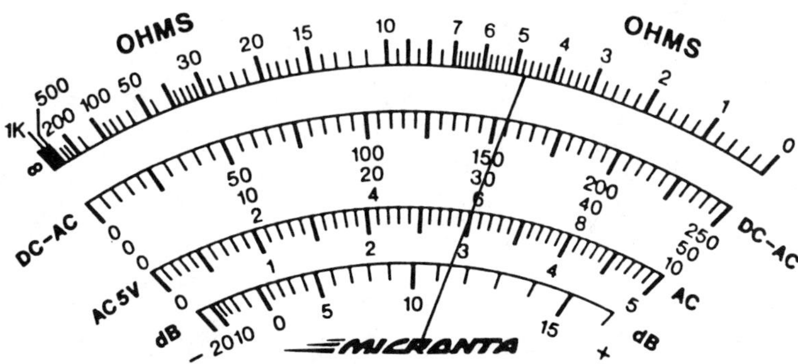

9. With meter set on DCV 2.5
the correct reading is

_______________________.

10. With meter set on DCV 2.5
the correct reading is

_______________________.

11. With meter set on DCV 2.5
the correct reading is

_______________________.

12. With meter set on DCV 2.5
    the correct reading is

    _______________________ .

SECTION VIII

Calculation Problems

1. State Ohm's law in 3 ways and define each symbol:

   a.

   b.

   c.

2. Using Ohm's law, if you have a 1.55 volt cell, and the resistance value of the coil is 2,000 ohms, give the current consumption in milliAmps.

   Answer________________.

3. Using Ohm's law, if you have a 1.55 volt cell, and the resistance value of the coil is 4,000 ohms, give the current consumption in microAmps.

   Answer________________.

4. Using Ohm's law, if you have a 1.35 volt cell, and the resistance value of the coil is 3,500 ohms, give the current consumption in microAmps.

   Answer________________.

5. If a circuit has a total resistance of 2,000 ohms and a current of 1mA is flowing through the circuit, what is the voltage of the battery?

   Answer________________.

6. If a current flow of 15μA is present in a 66,666 ohm circuit, what voltage is applied?

   Answer________________.

7. What voltage is required to sustain a 20μA current flow in a 40,000 ohm circuit?

   Answer________________.

8. If a 1.5 volt battery causes a 15mA current flow through the circuit, what is the resistance of the circuit?

   Answer________________.

9. 3μA of current is flowing in a circuit powered by a 1.5 volt cell; what is the resistance of the circuit?

   Answer________________.

10. To cause a 20mA current to flow in a circuit containing a 12 volt battery, what resistance will be required?

   Answer_________________.

SECTION IX

Practical Bench Questions

A.   Multiple Choice Questions
(Check the correct answer.)

1.   To ease the assembly of the train of wheels in a stepping motor watch, you could place the pillar plate on:

( )   a.   your movement holder
( )   b.   a piece of pitchwood
( )   c.   a steel bench block

2.   If a watch has two cells and one checks out OK and the other one is very low, you should:

( )   a.   definitely replace both cells
( )   b.   just replace the bad cell
( )   c.   just exchange the position of the cells so the good cell will be in the "active" cell-well.

3.   On many quartz analog watches, when the crown is pulled out to the "set" position, the watch will stop because:

( )   a.   the crown will rub on one's wrist and thereby stop the watch
( )   b.   the watch is designed with an "electronic shut-off" system, which also helps save the battery
( )   c.   when in setting position, the clutch wheel touches the stepping motor, causing it to stop

4.   On an LCD watch, if we push on a push button and nothing happens, the problem might be:

( )   a.   the push button was designed too short at the factory
( )   b.   the push button was designed too long at the factory
( )   c.   dirt between the push button and the case

5.   How much oil is recommended to be used on the outside magnetic area of the rotor in a quartz analog watch?

( )   a.   very little
( )   b.   none
( )   c.   just in 3 or 4 spots around the outside

6.   In quartz watches, the life of the battery is determined by:

( )   a.   the battery manufacturer predetermined this
( )   b.   the size of the battery is the only factor
( )   c.   the low current consumption and reduced friction of the watch

7. When cleaning an analog quartz watch:

    ( ) a. all parts should be cleaned in the ultrasonic cleaning machine
    ( ) b. no electronic parts should be cleaned in the ultrasonic machine
    ( ) c. no quartz analog watch should be cleaned in the ultrasonic cleaning machine

8. On an LCD watch, one of the polarizers fits over the display. Without it the watch would:

    ( ) a. not keep the correct time
    ( ) b. run, but we could not see the display
    ( ) c. not run

9. If an LCD watch comes in and one corner of the display has turned dark, you should:

    ( ) a. change the battery immediately
    ( ) b. replace the display
    ( ) c. clean the zebra strip

10. When numbers on an LCD appear dark against a white background:

    ( ) a. the "field effect" system is being used
    ( ) b. the LCD should be replaced
    ( ) c. the "dynamic scattering" system is being used

11. If the display (on an LCD watch) is blank after replacing the cell, then the first thing to try is:

    ( ) a. wait 15 minutes and see if watch will reset itself automatically
    ( ) b. short a reset mark, or tab marked AC (means all clear), to the positive side of the cell
    ( ) c. replace the display with a new one

12. To regulate a quartz watch that is designed to be regulated, it is necessary to place the watch on a timing machine designed for quartz and:

    ( ) a. turn the trimmer capacitor to the right if the watch is slow, and to the left if it is fast
    ( ) b. turn the trimmer capacitor in a clockwise direction while watching the results on a timing machine
    ( ) c. turn the trimmer capacitor in either direction while watching the results on a timing machine

13. The purpose of a zebra strip is to connect:

    ( ) a. the cell to the circuit
    ( ) b. the coil to the circuit
    ( ) c. the display to the circuit

14. Which of the following should be oiled in a stepping motor watch?

    ( ) a. the pivots of the train
    ( ) b. the coil pads
    ( ) c. the outside of the rotor

B.   True/False Questions

T   F

☐ ☐   1. Blocking the train of a quartz movement which does not have a load compensated circuit will cause the current consumption to approximately double.

☐ ☐   2. A quartz timer with an inductive pick-up must be used to time quartz watches with an inhibition or logic circuit.

☐ ☐   3. When we have an "open" coil, it can sometimes be repaired by using conductive epoxy neatly smeared over the broken area.

☐ ☐   4. A blocked train in a quartz analog watch will cause higher than normal current consumption.

☐ ☐   5. Light from your bench lamp can cause a circuit whose substrate is made of photo-sensitive film to have a higher than normal current reading.

☐ ☐   6. A "zebra" strip is never used to connect the circuit board to the LCD display.

☐ ☐   7. Gate time is the number of oscillations of the quartz crystal in a ten second period.

☐ ☐   8. The pulse period of a quartz analog watch is 5 minutes.

☐ ☐   9. When the reset switch is activated on a quartz analog watch, the watch will automatically be set exactly at midnight.

☐ ☐   10. All quartz watches using a stepping motor supply one pulse per second.

☐ ☐   11. Some integrated circuits can be damaged by static electricity.

☐ ☐   12. A defective or shorted circuit may cause a high current consumption.

☐ ☐   13. There is no difference between a high drain and a low drain cell.

☐ ☐   14. A watch with an advertised five-year cell has less friction in the train than a watch that has a guaranteed one-year cell.

☐ ☐   15. You can save the customer money by replacing only one cell when only one of the cells checks "bad" on the multimeter.

☐ ☐   16. An aging LCD panel may cause a high current consumption reading.

C.   Fill in the Blanks

1.   A load compensated circuit provides a second________________________if the rotor does not step when commanded to do so.

2.   A load compensated circuit allows the circuit of a quartz analog watch to operate at a lower rate of________________________ ________________________ during the normal operation.

3. A diagram of electronic circuitry showing electrical connections and identification of various components is known as a______________________.

4. An alkali metal used in the construction of batteries instead of silver or mercury is known as

______________________.

5. A (______________ ______________) circuit is a circuit in which the error of the quartz crystal frequency is corrected in the IC chip and the corrected rate is supplied to the coil.

6. Light from a bench lamp on a photo-sensitive substrate will cause______________________ current consumption.

7. A damaged coil can sometimes be repaired by painting the damaged area with ______________ ______________ epoxy.

8. Stepping motor quartz watches with second hands are required to "step"______________

______________.

9. A cell that is alkaline in nature and has no wet material is known as a______________ cell.

D. Short Answer Questions

1. Give the two most common reasons why a second hand of a stepping motor quartz (SMQ) should "jerk" every second but not progress.

a.

b.

2. When soldering on an electronic movement, what type of soldering iron and flux should be used?

3. What causes the rotor of a stepping motor quartz watch to continue turning in one direction only?

4. When a stepping motor quartz movement will not run below a voltage of 1.4V, does it indicate a problem in the Integrated Circuit, the coil, or the train?

5. How is it recommended that you clean the electronic parts of a quartz watch?

6. What is the recommended procedure to prevent damage from static electricity to a quartz watch you are working on?

7. How can you recognize a load compensated circuit?

8. Give 3 methods used to regulate quartz watches.

   a.

   b.

   c.

SECTION X

Answers to All Questions & Problems

SECTION II—Multiple Choice, p. 7

1.	A	45.	B
2.	C	46.	C
3.	B	47.	B
4.	A	48.	C
5.	B	49.	B
6.	A	50.	B
7.	B	51.	C
8.	B	52.	C
9.	A	53.	B
10.	A	54.	C
11.	A	55.	C
12.	A	56.	B
13.	B	57.	C
14.	A	58.	C
15.	A	59.	B
16.	A	60.	C
17.	B	61.	C
18.	A	62.	C
19.	B	63.	C
20.	B	64.	B
21.	B	65.	B
22.	A	66.	C
23.	C	67.	C
24.	B	68.	C
25.	B	69.	B
26.	A	70.	B
27.	C	71.	B
28.	C	72.	B
29.	A	73.	B
30.	B	74.	B
31.	C	75.	A
32.	C	76.	C
33.	B	77.	B
34.	B	78.	C
35.	A	79.	B
36.	C	80.	A
37.	C		
38.	C		
39.	B		
40.	B		
41.	C		
42.	A		
43.	C		
44.	A		

SECTION III—True/False, p. 17

1.	T	46.	F
2.	T	47.	T
3.	F	48.	T
4.	T	49.	T
5.	F	50.	T
6.	F	51.	T
7.	F	52.	T
8.	F	53.	F
9.	F	54.	T
10.	T	55.	T
11.	F	56.	F
12.	T	57.	T
13.	T	58.	T
14.	F	59.	T
15.	T	60.	T
16.	T	61.	T
17.	T	62.	F
18.	F	63.	F
19.	F	64.	F
20.	F	65.	F
21.	T	66.	F
22.	T	67.	T
23.	T	68.	F
24.	F	69.	T
25.	T	70.	T
26.	T	71.	T
27.	T	72.	T
28.	F	73.	T
29.	F	74.	T
30.	T	75.	T
31.	F	76.	T
32.	T	77.	F
33.	T	78.	T
34.	T	79.	F
35.	T	80.	T
36.	T	81.	T
37.	T	82.	F
38.	T	83.	F
39.	T	84.	T
40.	T	85.	F
41.	F	86.	T
42.	T	87.	F
43.	T	88.	T
44.	F	89.	T
45.	F		

SECTION IV—Matching Questions, p. 21

1.	A	16.	C
2.	B	17.	A
3.	C	18.	B
4.	B	19.	A
5.	A	20.	B
6.	C	21.	C
7.	A	22.	C
8.	B	23.	E
9.	C	24.	F
10.	F	25.	D
11.	E	26.	C
12.	D	27.	A
13.	B	28.	B
14.	A	29.	C
15.	C	30.	E

SECTION V—Completion Questions, p. 24

1. ohms
2. Ampere
3. volt
4. voltage
5. movement, flow and propagation
6. diode
7. capacitor
8. hertz
9. ferrous
10. parallel
11. series
12. transistor
13. voltage
14. 4.5
15. 1.5
16. repel, attract
17. reverse
18. negative, positive or cathode, anode
19. 32,768
20. Amps
21. ohms
22. volts
23. carbon
24. AC
25. two
26. short circuit

SECTION V (continued)

27. conductor
28. insulated
29. conductor
30. circuit
31. resistance
32. magnetic field
33. silver
34. permanent magnets
35. current flow, less resistance, weaker magnetic field
36. polarity
37. electrons, current
38. voltage
39. nonconductor or insulator
40. resistor
41. capacitor
42. mechanical
43. three
44. transistor
45. trimmer capacitor
46. electrons
47. shorted
48. voltage
49. cell, battery
50. permanent magnets
51. electromagnets
52. voltage
53. resistance
54. Amperage
55. electric
56. sparking or arcing
57. cell, battery
58. LCD
59. LED
60. negative, positive or − to +
61. resistance
62. base

SECTION VI—Short Answer Questions, p. 27

1. a. permanent magnet
    b. electromagnet
    c. temporary magnet
2. the direction of current flow
3. repel each other
4. number of windings of the coil
5. temporary or electromagnet
6. maintains magnetic properties almost indefinitely
7. has magnetic properties only under the influence of another magnet
8. has magnetic properties only when voltage is applied, energized coil
9. concentrates and increase the magnetic field without increasing size of coil
10. 1.55

11. 1.5
12. to increase voltage
13. no
14. reverse leads
15. 10 mA
16. 15 mA
17. .2 mA
18. 2,200 Ω
19. ohms
20. new batteries
21. no
22. rotor

SECTION VII—Meter Reading Questions, p. 29

1. 30 Ω
2. 150 Ω
3. 1,000 Ω
4. 24,000 Ω
5. 16 μA
6. 34 μA
7. 48 μA
8. 2 μA
9. 1.55 V
10. .85 V
11. 2.25 V
12. .2 V

SECTION VIII—Calculation Problems, p. 33

1. a. E = 1 X R,   E = volts
    b. 1 = E ÷ R,   I = amps
    c. R = E ÷ I,   R = ohms
2. .775 mA
3. 387.5 μA
4. 385.7 μA
5. 2 volts
6. 1 volt
7. .8 volt
8. 100 ohms
9. 50,000 ohms
10. 600 ohms

SECTION IX—Practical Bench Questions, p. 35

Section A		Section B	
1.	c	1.	T
2.	a	2.	T
3.	b	3.	T
4.	c	4.	T
5.	b	5.	T
6.	c	6.	F
7.	b	7.	F
8.	b	8.	F
9.	b	9.	F
10.	a	10.	F
11.	b	11.	T
12.	c	12.	T
13.	c	13.	F
14.	a	14.	T
		15.	F
		16.	T

SECTION IX (continued)

Section C

1. impulse
2. current consumption
3. schematic
4. lithium
5. load consumption
6. high or higher
7. conductive
8. once a second
9. lithium

Section D

1. a. a blocked train of wheels
   b. low battery voltage
2. a low voltage cordless soldering iron should be used, and no flux, or a noncorrosive flux
3. alternatingly reversing the polarity of the coil current
4. a problem in the train
5. with a cotton swab and alcohol
6. touch to bench lamp to ground yourself
7. by the technical guide or by blocking the train of wheels
8. a. trimmer capacitor
   b. switch
   c. breaking conductors
   d. adding/removing graphite
   e. screw down contact

SECTION XI

Informative Questions & Answers

1. **What four factors affect the flow of current through a wire?**

    Length, thickness, composition and temperature; therefore, 2 wires of the same length, thickness, composition and temperature would have the same resistance as well as the same conductance.

2. **What is an insulator used for?**

    To prevent or block the flow of current at a specific place, such as using an insulator under a cell to prevent shorting the cell.

3. **What is the voltage of mercury cells?**

    1.35 volts

4. **What happens when you reverse the flow of current in an insulated coil?**

    Polarity of the coil is reversed.

5. **What are temporary magnets made of?**

    Soft iron, or similar materials.

6. **How is an energy cell tested?**

    With the cell in the watch, the voltmeter is set on the appropriate scale. The negative lead of the meter is touched to the negative battery contact while the positive lead of the meter is touched to the positive battery contact.

7. **What should a watchmaker do when an electric/electronic watch is brought in for repair?**

    First, he should check the cell, and then should attempt to separate the mechanical problems from the electronic problems.

8. **What are the major advantages of the LCD watch over the LED watch?**

    The display of the LCD watch is on continuously, and the current consumption is several times less than the LED watch.

9. **What is the purpose of an iron core in an electromagnet?**

To increase, concentrate, and direct the magnetic field of force.

10. **What is an electromagnet?**

A magnet composed of a coil of insulated wire which when a current is passed through it causes the coil to have a north and south magnetic pole. This makes an electromagnet a magnet only when voltage is applied; thus, it is a temporary magnet.

11. **What is a battery?**

A battery is composed of two or more cells, although usage of the term has come to denote a single cell.

12. **What causes "sparking"?**

Sparking is caused by a backwash of negative current in a circuit when a mechanical switch is opened. This causes a carbon buildup as well as pitting of the contacts. In fact, this pitting was the major problem with the electric balance wheel watch.

13. **What is a trimmer capacitor?**

A trimmer capacitor is a variable capacitor used in the electronic watch to vary the frequency of the quartz crystal, thus regulating the watch.

14. **What would be the result if a trimmer capacitor became disconnected from the circuit?**

If the trimmer capacitor should become disconnected from the circuit, the frequency of the quartz crystal would increase slightly. If the watch were placed on the timing machine and the trimmer capacitor were turned you would see no change in frequency.

15. **When would you use a high drain silver cell as opposed to a low drain silver cell?**

If the cell of a watch was missing and you had no way of knowing if the watch required a high drain or a low drain cell, you would be safe in using a high drain cell with dimensions that would fit the battery cavity. However, if a watch has any of the following functions, it would require a high drain cell: alarm, night light, etc.

16. **Which part of the stepping motor quartz consumes the most current?**

The stepping motor. The oscillating circuit and the dividing circuit consume very small amounts of current.

17. **How would the timekeeping element of a watch be affected if the mechanical parts were removed?**

The cell, the oscillator circuit, and the divider circuit would be unaffected by the removal of the mechanical portion of a watch. However, the mechanical parts are needed for the display of time.

18. What are nonconductors?

They are materials that allow little or no current to pass through. Some examples are: air, distilled water, glass, wood, paper, bakelite, plaster. Some materials, such as tap water, offer a lot of resistance to current flow, but allow a small amount to pass.

19. What occurs when electric current flows through a wire conductor?

When current flows through a wire conductor, a magnetic field is created around the wire, forming a magnetic north and south pole.

20. What happens when more voltage is applied to a conductor?

When more voltage is applied to a conductor, there is more current flow. When excessive voltage is applied, damage/overheating could occur.

21. What is the difference between DC current and AC current?

DC current flows in one direction only. AC current flows from negative to positive, and positive to negative, alternately, at a set frequency.

22. If the insulation of a coil is damaged, causing several coils to touch, what would be the results?

The resistance of this coil would be less because the effective length of the wire has been shortened; therefore, more current would flow. The magnetic field would be weaker (smaller) also.

23. What are the capabilities of a capacitor?

The capacitor has the capability of storing a predetermined amount of energy (somewhat like a battery). By doing this it has the ability to control and stabilize voltage. A capacitor can discharge sudden surges of current, unlike a battery.

24. How does heat affect resistance through a conductor?

When current flows through a conductor which has resistance, a small amount of heat is generated because of the resistance to the flow of current. Therefore, some energy is lost through heat.

25. When does current flow through a diode?

Since electrons flow from negative to positive, when a diode is placed in a circuit the negative side of the diode must be connected to the negative side of the circuit, and the positive side of the diode must be connected to the positive side of the circuit in order for electrons to flow through the diode. If the diode were reversed, no electrons would flow.

26. **How is the base of a transistor biased?**

Often a resistor is used to connect a transistor base to the power source, thereby supplying a small amount of current to the transistor base to "turn it on." Without this current, the transistor will not work.

27. **How many times must the divider circuit divide the standard quartz crystal frequency (32,768 hz) by two to reach one hertz?**

In this divider circuit is a series of 15 flip-flops.

28. **When measuring the resistance in a coil that is shorted, what meter indication would you get?**

The needle would swing to zero on the ohms scale.

29. **How does a capacitor give off a more sudden surge of energy than the power source?**

A capacitor can be charged to the same voltage as its power source but has the capability of being discharged instantly thereby giving off a more sudden surge than its power source.

30. **Do all resistors completely block the flow of current?**

No, resistors are designed to restrict the flow of current and allow a certain predetermined amount to flow.

31. **What is the function of a stator in an electromagnet of a quartz analog watch?**

The stator acts as a core of the electromagnet, although it is not completely in the center of the coil.

32. **Does a permanent magnet have influence on an electromagnet?**

Yes. Just as two permanent magnets would (like poles repel, unlike poles attract), the further the distance between them the less they will influence each other.

33. **What happens when a magnet is passed over a coil of insulated wire or a conductor?**

When a magnet is passed over a coil of insulated wire or a conductor the magnetic field cuts through the wires, creating a flow of current in one direction. If the direction of the magnetic field is reversed, current in the coil will also be reversed.

34. **Why are quartz timepieces more accurate than balance wheel timepieces?**

The quartz crystal oscillator is virtually unaffected by outside influences, such as shock, temperature changes, friction, lubricant failure, etc., which do affect the balance wheel watch. The oscillation frequency of the quartz crystal is also much more stable than the balance complete.

35. **What is a booster coil, and what is its purpose?**

The booster coil is an additional coil placed on the circuit board to increase voltage to the alarm circuit. This increase is necessary because 1½ volts is insufficient to power the alarm.

36. **What is a "wet" cell?**

A "wet" cell is composed of a container holding a fluid that acts as an electrolyte, and has two dissimilar metals inserted which form the cathode and anode. The wet cell was invented by Allesandro Volta.

37. **Which part of the multimeter is used to measure resistance?**

The ohmmeter.

38. **Which part of the multimeter is used to measure current?**

The ammeter.

39. **What is a multimeter?**

The term "multimeter" refers to two or more meters, sharing the same housing. Most multimeters contain an AC voltmeter for measuring alternating current voltage (such as house current); a DC voltmeter for measuring direct current voltage (such as watch cells); an ammeter for measuring current consumption (measured in Amps, milliAmps, and micro-Amps); and an ohmmeter for measuring resistance in ohms (such as coils).

40. **What are some conductors?**

In order of superiority, the following are good conductors: silver, copper, gold, aluminum, zinc, tungsten, brass, tin, iron, nickel, platinum, soft steel, mercury, and cast iron. These conductors allow free electron flow.

41. **How many chambers does a diode contain?**

A diode is built like a transistor except that a diode contains two chambers, and a transistor contains three.

42. **What occurs when you have a short circuit?**

The resistance will be reduced considerably; therefore, more current would flow.

43. **What would be the result of adding a resistor in a series circuit?**

There would be less current flow, and also a higher resistance reading.

44. **What would be the result if a nonconductor or insulator were placed in a circuit?**

There would be no current flow.

45. What is an electrical shunt used for?

To collect and redirect current flow or to bypass its normal route by using certain kinds of metals.

46. How does one measure resistance?

With the ohmmeter.

47. What is the most commonly used frequency of quartz crystals in watches?

32,768 hertz.

48. What are two disadvantages of the LED watch?

One: a button has to be depressed to activate the display. Two: it consumes large amounts of current.

49. Which type of watch uses a mechanical switch?

The electric balance wheel watch is one that uses a mechanical switch. The addition of an electronic switch (transistor) makes the watch an electronic watch.

50. What is the difference between cells connected in series and cells connected in parallel?

When like cells are connected in parallel, the voltage will remain the same. When like cells are connected in series, the voltage will be the sum of their voltages.

51. What is the difference between a large cell and a small cell with the same voltage and chemical composition?

The larger cell would have a greater milliAmp hour capacity.

52. What would the reading on the ohmmeter be if a coil or circuit was "open"?

If either the coil or circuit was "open" the needle of the ohmmeter would not move from infinity. An "open" circuit or coil is one in which there is a break so current cannot flow.

53. Do all magnets have polarity?

Yes. A permanent magnet has polarity constantly, a temporary magnet has polarity when under the influence of another magnet, and an electromagnet has polarity when current is applied.

54. Should any electronic or electric watch be demagnetized?

No. To demagnetize any electric or electronic watch would destroy the permanent magnets.

55. **What would the cell in an electric or electronic watch compare with in a mechanical watch?**

The mainspring in a mechanical watch provides power to drive the watch, as the cell does in an electric/electronic watch.

56. **If a balance complete has a permanent magnet attached to the rim and is pulsed with energy from an electromagnet, could it be used to keep time?**

Yes. In actual fact, this is the principle upon which the first electric watches were based. The only addition to the above was a mechanical switch to turn the electromagnet on and off at the proper time to cause the balance to swing with a regular motion. The mechanical switch was eventually replaced with a transistor making the electric wheel watch into an electronic balance wheel watch.

57. **Name three kinds of magnets.**

Permanent, temporary, electromagnet.

58. **What is the law of magnets?**

Opposite poles attract and like poles repel. Thus, if a north and south end of two magnets are brought together, they will attract each other; and if two north ends are brought together, they will repel each other.

59. **How would hitting a magnet affect it?**

To jar or to hit a magnet could temporarily excite the atoms, causing a loss of magnetism.

60. **What will make an electromagnet stronger?**

More coils of insulated wire or a soft iron core.

61. **What happens when voltage is applied to a quartz crystal?**

When voltage is applied to a quartz crystal, the crystal will vibrate or oscillate at a frequency depending on the way in which the crystal is cut. This is known as the piezo-electric effect.

62. **What would be the difference in the results of connecting a 1.55 volt cell and a 1.35 volt cell to resistors of equal value?**

If the resistance remains the same and the voltage is decreased, the current flow will also be decreased (Ohm's Law $I = E \div R$). Therefore, fewer electrons will flow.

63. **What is the composition of a magnetic field?**

A magnetic field is believed to be a flow of electrons in a circular pattern, perpendicular to the axis of the conductor through which the current is flowing.

64. What are permanent magnets made of?

Permanent magnets are usually made of steel, samarium cobalt, or other alloys with like characteristics. Some examples of permanent magnets are: horseshoe magnet, bar magnet, compass needles, radio speaker cone magnets, rotors of stepping motor watches, etc.

65. What materials will block magnetic lines of force?

Very few materials will block the magnetic lines of force. Only lead and some other dense elements can block these lines of force. All metals, whether ferrous or nonferrous, will allow magnetic lines of force to flow through them.

66. What is current?

Current is movement, flow, or propagation of electrons through a conductor.

67. What is voltage?

Voltage is the electromotive force that causes current to flow. It can be considered to be electrical pressure.

68. How do you read current consumption?

The amount of current which is being consumed can be measured as it flows, using an ammeter connected in series.

69. What is required for current to flow?

In order for current to flow, there must be a source of voltage and a circuit for the current to flow through.

70. When measuring DC voltage or current consumption, what does it mean if the needle registers a negative reading?

This means that the meter probes are reversed and you should exchange (reverse) the positions of the probes.

71. What is the meaning of milliAmp and microAmp?

"Milli" means thousandth, and "Micro" means millionth. Therefore, one milliAmp is one thousandth of one Ampere, and one microAmp is one millionth of one Ampere. One Amp = 1,000 milliAmp = 1,000,000 microAmp. The current consumption of modern quartz watches is measured in microAmps. Abbreviations: microAmp—μA; milliAmp—mA.

72. How is an ohmmeter used?

The ohmmeter, having its own internal power source to power the meter, is capable of measuring the resistance of any substance or object. When using the ohmmeter to measure the value of a resistor, it does not matter which probe is placed where because current will flow in either direction through a resistor. CAUTION: Power source must be disconnected first.

73. How does one "zero" an ohmmeter?

Before using your ohmmeter, it should be zeroed. This is done by setting the function switch to the resistance scale you wish to use. Then place the probe tips together. This should cause the needle to come to rest on the zero mark of the resistance scale of the dial. If the needle does not come to rest at the zero, you should be able to move the needle to zero by turning the zero adjust knob. If this cannot be accomplished by turning the zero adjust knob, you will have to replace the battery inside the multimeter.

74. How can a coil be checked if it cannot be separated from the circuit board?

To test this type of coil you should supply less than .3 volts from a variable power supply, set your function switch to read microAmperes, remove the cell from the watch, place your probes on the coil connection pads, and read the current consumption on your meter. When checking a coil in this manner you cannot read the resistance but if you can read current flow through the coil it proves that the coil is not open.

75. How does one check the output of the circuit of a stepping motor watch?

First, determine that the cell and cell contacts are good. Then with the cell in place, set your meter function switch to the lowest DC volts scale and place the meter probes on the coil connection pads. Your meter needle will swing alternately upscale and downscale in sync with impulses from the circuit to the coil.

76. How may one keep the rotor upright while reassembling the train of wheels?

During reassembly, if you place the pillar plate on your steel bench block the rotor magnet will be attracted to the steel block and will hold itself in an upright position.

77. How does one check a stepping motor for train freedom?

Since the rotor has a tendency to stop at each half turn it is impossible to check train freedom in the conventional manner. Train freedom can only be checked by lowering the input voltage until the watch stops stepping; then raise the voltage slowly until it begins to step again. The watch should run at 80% of normal cell voltage. Example: 1.5 volts x .80 = 1.2 volts.

78. What would be the effect of placing a capacitor in parallel with the power supply while checking current consumption of a stepping motor watch?

A capacitor placed in parallel with the power source will tend to provide a more constant flow of electrons; thus, the meter will tend to give a more constant reading and the needle will not swing as far upscale at impulse as it does without the capacitor.

79. **How does one determine if a coil is defective when it is separated from the circuit?**

Once a coil is separated from the circuit board, you can read its resistance with an ohmmeter. Coils will usually read between 1,000 ohms and 4,000 ohms. Should a coil read below 1,000 ohms or above 4,000 ohms, you can consider the coil to be shorted (low resistance reading) or open (high resistance reading). In order to determine if the coil winding is grounded to the metal core, you would place one meter probe on the metal core and the other probe on one of the coil connection pads. If the ohmmeter indicates any resistance is present, the coil is grounded to the core.

80. **What would be the result of putting a high drain cell in a watch that requires only a low drain cell and vice versa?**

If you inadvertently placed a high drain cell in a watch which only requires a low drain cell, the watch would run normally but the cell may not power the watch quite as long. If you were to place a low drain cell in a watch requiring a high drain cell, the watch would keep time but the alarm, night light, and any other high drain feature may not work.

81. **What is a "zebra" strip in an LCD digital watch?**

The zebra is the elastomer connector made of alternating insulating and conductive strips connecting the circuit board to the conductive contacts of the LCD display panel.

82. **What is the difference between a high drain and a low drain cell?**

The high drain cell can furnish a large number of electrons upon demand to operate the extra functions of a watch (such as night light, alarm etc.), in addition to operating the timekeeping element. The low drain cell is designed to furnish a limited number of electrons and should be used only to power the timekeeping element when no additional functions are present.

83. **What are some common causes of high current consumption?**
1. Battery leakage (chemical contamination).
2. Moisture.
3. Defective or shorted circuit.
4. Bench lamp on a photo-sensitive substrate.
5. Blocked train.

84. **Briefly describe a load compensated circuit.**

A load compensated circuit is one in which a second, longer voltage pulse is generated in the event the rotor does not rotate the number of degrees it was designed to step as a result of the normal pulse.

85. **What is the main advantage of a load compensated circuit?**

It allows the movement to operate at a lower rate of current consumption during normal operation.

86.   Why are some movements advertised as having a five-year battery life?

The low current consumption due to the increased efficiency of the circuit design, and less friction in the train, coupled with the current capacity of the battery provides a predicted cell life of five years or more.

87.   What is an inhibition or logic circuit in a quartz watch?

It is a circuit that is designed to correct within the IC chip a quartz crystal frequency that is near but not at the exact required frequency. The corrected rate is supplied to the coil. A trimmer capacitor on this watch is unnecessary, and if one is present, it would be ineffective.

88.   Why must a timer with an inductive pickup be used to time a quartz watch which has a logic or inhibition circuit?

First, let's review the makeup of a quartz analog watch. The cell provides power to a part of the integrated circuit which changes the direct current from the cell to alternating current. The output of this part of the circuit is used to cause the quartz crystal to oscillate. The oscillations are returned to the circuit at a frequency of 32,768 hertz. This frequency is fed into the divider chain contained in the integrated circuit, and emerges as one pulse per second. This output is fed to the coil. The pulses from the coil cause the stepping motor to turn.

The inductive pickup of a timing machine receives the pulse supplied to the coil that drives the step motor. The timer measures the frequency of these impulses and gives a readout of the average of these measured impulses.

The acoustic pickup of a timing machine receives and reads the oscillations of the quartz crystal. Therefore, if you were to attempt to read the frequency of the quartz crystal in a watch which has an inhibition or logic circuit with an acoustic pickup, you may not get an exact reading of 32,768 hertz. This is due to the almost constant corrections being made by the inhibition circuit.

89.   Some quartz circuit designs use a photo-sensitive film as a substrate. How does this affect testing procedures?

Your bench lamp may generate a high current reading because of the light-sensitive characteristic of the substrate. To obtain a correct reading, turn off the bench lamp, or cover the circuit with opaque material.

90.   What is gate time?

Gate time is the sampling period during which the period between two or more output pulses of the IC are averaged, then converted to a daily or monthly rate.

91.   What is the pulse period of a quartz analog watch?

It is the time between the starting point of one pulse and the starting point of the next consecutive pulse.

92.   **Briefly describe the function of the seconds-setting, or reset switch.**

When the stem is pulled out to setting position, the reset switch contacts a point on the circuit board, causing the pulses to the coil to stop. When the stem is pushed in, the watch will start exactly one second later, allowing very accurate time setting. In most cases there is also a lever that blocks one of the train wheels so that the hands can be set without spinning the train.

93.   **Why is a quartz crystal aged?**

A quartz crystal, when used immediately after it is cut, does not have a steady frequency. Its true frequency is reached only after it has been set aside for a certain length of time. This is called aging.

94.   **Why do numerals appearing on a "field effect" liquid crystal display appear dark against a white background?**

The liquid crystal sandwiched between two glass sheets has its molecules twisted 90 degrees between the top and bottom sheets. This is accomplished by a special treatment to the glass sheets. Two polarized sheets are also arranged to be staggered by 90 degrees from each other with reference to their polarization axis. Without one of these sheets, no display would be visible.

When a voltage is applied between a segment electrode and the common electrode, the resulting electric field acts to orient the liquid crystal molecules under its influence vertically. Thus, the admitting waves directly pass through them without being twisted, and are prevented by the lower polarization sheet from reaching the reflecting mirror.

Accordingly, the voltage of the field-affected segment part is blacked out. That is, the segment is displayed in contrast with the surrounding nonaffected parts. In practice, seven segments on the upper plate glass are applied with a voltage selectively to form any desired numeral pattern.

95.   **Why should you replace both cells in a watch if only one of them is "dead"?**

If one cell in a watch is found to be below its normal cell voltage, it is safe to assume that the other cell is nearing the end of its life also. Since there is no test to determine the remaining current capacity of a cell, the only way to keep your customer happy is to replace both cells. Another factor to consider: placing a new cell with a full current capacity into a circuit with a cell nearing the end of its life will cause the new cell to discharge a part of its current capacity into the cell nearing the end of its life—thus making the current capacity of both cells nearly equal after a short period of time.

96.   **Why is there sometimes no display on an LCD watch after the cell has been replaced?**

Some LCD watch circuits are designed with a reset, or "all clear" function that will clear the memory of the circuit. The absence of display is a reminder that the voltage supply has been interrupted and that the reset, or all clear function, must be activated before the display will be restored. The reset or AC pads on the circuit board are shorted to the positive side of the cell to activate the reset or all clear function.

97. What may cause a push button on an LCD watch to be inoperable?

In most cases you will find that dirt has built up between the push button and the case and will not allow the button to move. You may also encounter dirt or corrosion on the contact that the push button is supposed to activate.

98. What would be the result of oiling the outside of the rotor in a stepping motor watch?

Since that rotor rotates between the poles of the stator and there is very little clearance between the rotor and the stator, the oil would cause a great resistance to the turning of the rotor. This in turn would cause a rise in current consumption as well as possibly cause the rotor not to step at each impulse of the coil. Only the setting parts and train wheel pivots should be lightly lubricated. This includes the stepping motor pivots.

99. Can an open coil of a quartz analog watch be repaired?

If you can see the damaged area of the coil, and not too many turns of the coil are cut, the coil can be repaired using a conductive epoxy or conductive paint. Care should be exercised to cover the damaged area with the liquid conductive mixture, but not to allow it to touch the two pads which attach the coil to the circuit board. It is best to apply the liquid while the coil is attached to the ohmmeter. This way you can know when the coil is repaired because you will get a resistance reading on the ohmmeter.

100. How do you determine in which direction to turn a trimmer capacitor to regulate a quartz watch?

In nearly all cases, there is no way to determine beforehand whether the rate will increase or decrease by turning the trimmer capacitor in either the left or the right direction. The best method to use in regulating a quartz watch is to place it on a timing machine and observe the reading as you turn the trimmer capacitor in either direction.

101. Should the electronic parts of a quartz analog watch be cleaned in an ultrasonic cleaning machine?

No. The electronic parts should never be cleaned in an ultrasonic cleaner. If you deem it necessary to clean the electronic parts you may use a cotton swab moistened in alcohol.

102. Can integrated circuits, such as those used in quartz watches, be damaged by static electricity?

Any integrated circuit can be damaged by static electricity. The circuits in the earlier electronic watches are more susceptible to damage than the circuits in the newer models. However, all are subject to damage unless precautions are taken. The circuit is insulated against damage from static electricity while inside the closed case, but when the case is open, the circuit is exposed. To prevent damage from static electricity, never touch the circuit without first touching your bench lamp, or some other grounded object, to drain off the static electricity from your body.

103. Do all stepping motor watches step once a second?

No. Generally, only a watch having a second hand is required to step once each second. Watches without second hands are designed to step at various frequencies, from six seconds up to one minute, thereby greatly lowering current consumption and extending cell life.

104. What is a schematic?

A diagram of electronic circuitry which shows electrical connections and the identification and location of various components, such as integrated circuits, discrete components, quartz crystals, cells, etc. is called a schematic.

105. What is different about the composition of the lithium cell?

Lithium is a silvery-white metal, slightly harder than sodium, but much softer than lead. It is the lightest of all metals. Lithium cells use highly reactive alkali metals and contain no water or wet material, unlike silver oxide or mercury cells. The components of the lithium cell do not react with one another unless a load is placed across their electrodes; therefore, their shelf life is virtually limitless.

106. Why does the second hand on a quartz analog watch "jerk" each second but not progress?

In all quartz analog watches this symptom can be caused by a blocked train. In some earlier watches this was an indication of low battery voltage, but in later models there are other low voltage indicators such as the second hand advancing in two-second increments at two-second intervals, etc.

107. What causes the stepping motor to turn in one direction only?

Electrically, alternatingly reversing the polarity of the coil current, coupled with the mechanical placement of the stator in relation to the rotor. These cause the rotor to turn in one direction only.

108. How important is the soldering iron and flux used to solder circuit boards in watches?

When soldering watch circuit boards, two precautions are imperative. First, the soldering iron to be used must be of the cordless type and must have a tip that is electrically insulated from the voltage source, such as the "Isotip" soldering iron. Second, and of equal importance, is the solder used. Solder without flux is ideal because all fluxes cause corrosion to some extent. In the event that a flux must be used, a rosin flux would be the least corrosive.

109.   Why would a stepping motor watch step at 1.5 volts and refuse to step at 1.4 volts?

This would be an indication of lack of train freedom. If there is sufficient train freedom, a watch should run at 80% of normal cell voltage. Example: 1.5 volts x .80 = 1.2 volts.

110.   What can cause a liquid crystal display panel to become defective if there has been no physical damage?

Two things that can cause a liquid crystal display panel to become defective are: 1) The aging of impurities that were originally in the liquid. 2) Leakage within the display panel which allows impurities to contaminate the liquid. Both the aging impurities and the leakage can be detected as a higher than normal current consumption.

111.   Is there a way to determine if a stepping motor watch has a load compensated circuit?

If the quartz stepping motor watch has a second hand, you can determine whether it is a load compensated circuit by blocking the train of wheels while measuring current consumption. If a load compensated circuit is incorporated in the watch, the current consumption will rise to between 6 and 8 microAmps, which is several times its normal consumption.

In the case of other quartz stepping motor watches with second hands, but without a load compensated circuit, a blocked train of wheels will cause a rise in current consumption to approximately twice its normal rate.

If you are blocking the train of wheels to determine if a load compensated circuit is present, be aware that some of the older quartz analog watches have a normal current consumption of 4 to 5 microAmps. With the train of wheels blocked and the current consumption doubled, you would get a current consumption reading consistent with a load compensated circuit, although it does not have one. For this reason, the best way to determine if a watch has a load compensated circuit is to check the technical guide.

In the case of the stepping motor watch which has no second hand (one that does not step once each second), a blocked train of wheels will not show a rise in current consumption that can be read on an analog type multimeter.

112.   What does it mean if one area of the display on an LCD watch has turned dark?

A darkened area of a display means that a fracture has occurred in the seal of the display, allowing contamination of the liquid. This problem will only get progressively worse; therefore, the display panel must be replaced.

113.   What are some methods of regulating a quartz watch?

1.   Trimmer capacitor.
2.   Switch.
3.   Breaking conductors on a circuit board.
4.   Adding or removing graphite (pencil lines) from a fixed capacitor.
5.   Screw down a contact to a positive or negative or neutral point on a circuit board.

The trimmer capacitor is the method most often used. One manufacturer used a pair of rotary switches which looked much like the trimmer capacitor. One was used for coarse regulation and the other for fine regulation. One manufacturer recommended breaking certain conductors on the circuit board to reduce capacitance, causing a gain in time. One manufacturer recommended adding or removing graphite from a field capacitor, causing it to gain or lose. One manufacturer used screws to make contact with a positive, negative, or neutral point on a circuit board.

The last four methods listed above are for specific models or calibers. They should not be attempted on every movement. It is best to obtain a technical guide for specific instructions on which models or calibers require which method.

114. How should watch batteries be disposed of?

Miniature batteries, such as we use in watches, pose a danger to small children and to the environment.

If a child swallows a battery, your private physician or emergency room physician can call the National Capital Poison Center collect for treatment advice. Information can also be obtained from the National Button Battery Ingestion Hotline at (202) 625-3333. Both these agencies are available 24 hours a day, 7 days a week.

Heavy metals such as mercury found in the miniature batteries also pose a threat to the environment.

For safe disposal of used batteries, return them to a member of AWI, or send them to AWI Central, 3700 Harrison Avenue, P.O. Box 11011, Cincinnati, Ohio 45211.

SECTION XII

Recommended Reading

RECOMMENDED BOOKS & TECHNICAL BULLETINS

Repairing Quartz Watches
     by Henry B. Fried

Digital Watch Repair Manual
     by Louis A. Zanoni

The Quartz Watch Repair Manual
     by Louis A. Zanoni

Swiss Technical Bulletin #39—Repair of ESA Electronic Calibers with Quartz-Crystal Resonators

Technical Bulletin—General Instructions, Seiko Analog Quartz

Technical Bulletin—General Instructions, Seiko Digital Quartz

ARTICLES IN HOROLOGICAL TIMES

(Taken from 10-Year Index)

BUTTON BATTERIES

Every year in the United States hundreds of people of all ages accidentally swallow miniature disc or "button" batteries of the type used to power hearing aids, watches, and calculators. The National Capital Poison Center in Washington, D.C. has been studying what happens when one of these button batteries is swallowed. Most button batteries pass through the body and are eliminated in the stool. However, sometimes batteries can get "hung up," and these are the ones that cause problems. A battery that doesn't move through the body may adhere to tissue, leak or break open, and chemical burns or poisoning may be the result. When a battery is swallowed, it is impossible to know whether it will pass through or get "hung up." If anyone swallows a battery, this is what you should do:

1. Call your poison center or the 24-hour National Button Battery Ingestion Hotline at (202) 625-3333 (TTY 202-625-6070) **immediately**.

2. If readily available, provide the battery identification number (from the package or from a matching battery).

3. An X-ray must be obtained right away to be sure that the battery has gone through the esophagus into the stomach. (If the battery remains in the esophagus, it must be removed immediately. However, most batteries move on to the stomach and can be allowed to pass by themselves.)

4. Watch for fever, abdominal pain, vomiting, or blood in the stools. Report these symptoms immediately.

6. Watch the stools until the battery has passed. Clean the battery, tape it to a card, or wrap it carefully and mail it in an envelope to:

NATIONAL CAPITAL POISON CENTER
Georgetown University Hospital
3800 Reservoir Road, N.W.
Washington, D.C. 20007

Be sure to include your name, address, and telephone number.

Your private physician or emergency room physician may call the National Capital Poison Center collect for treatment advice when a battery has been swallowed. We are on duty 24 hours a day, 7 days a week.

Button batteries may also cause injury when they are placed in the nose or in the ears. Young children and the elderly have been particularly involved in this kind of accident. Symptoms to watch for are pain and/or a discharge from the nose or ears. **Do not** use nose or ear drops until the person has been examined by a physician, as these fluids can cause additional injury should a battery be involved.

THE SIMPLE, ONE SWITCH POWER SUPPLY

By Buddy Carpenter, CMC, CEWS

For some years now a number of watchmakers have been "bugging" us to come up with a variable voltage power supply to be used with a multimeter to make all the tests necessary to repair all electronic watches. The catch is that they want only one switch to perform all the functions. Of course, we told them it could not be done. One of them, Jim Broughton by name, would not take "no" for an answer. So, at his insistance, here is our answer.

To make this variable voltage power supply attachment you will need the following items:

1. One LM324 Quad Op Amp   276-1711
2. One 9 Volt Transistor Battery
3. One 9 Volt Battery Connector   270-325
4. One 5000 Ohm Potentiometer   271-1714
5. One 220 MFD Electrolytic Capacitor   272-1017
6. Two Micro Test Clips   270-370
7. One 2 Position Push Terminal   274-315
8. One Economy Case   270-222
9. One 14 Pin Socket   276-1999
10. One ALCO, MTA406P four pole, double throw switch with open center

(Items 1 through 9 can be found at Radio Shack by these catalog numbers. Item #10 can be found at any other electronic supply house.)

These items can be used to convert the adapter on a Citizen meter or can be built into a small apparatus box from Radio Shack or similar electronics supplier and used with any multimeter. To convert a Citizen meter adapter, you need only remove the switch, battery connectors, the 3 jacks that the meter leads plug into, the capacitor, and all wiring except the wires connected to the positive and negative meter plugs.

The placement of components is not critical except in the case of the Citizen adapter. In the Citizen adapter, the potentiometer (pot.) must be mounted through the right side of the box, and the switch must be mounted through the front of the box (see Figure 3). If making an adapter for other multimeters you need only take care to keep the pot., switch, and push connectors near the edge of the lid so as not to interfere with the battery placement. All components can be mounted on the box lid to make assembly and wiring easier. Be sure to mount the switch so that terminal #1 is located in the upper left as you view it from the inside of the box lid (see Figure 2).

1.   To begin wiring, start by soldering an insulated wire between the following terminals on the back of the switch: 3 to 8, 12 to 10, 4 to 7 to 6.

2.   The capacitor can be soldered between terminals 5 and 9, being careful to place the negative lead of the capacitor on terminal 5.

3.   Solder the red lead to the battery connector to terminal 11 of the switch. Also solder a wire from terminal 11 of the switch to terminal 4 of the I.C.

4.   Solder any insulated wire between terminal 10 of the switch and terminal 1 of the pot.

5.   Solder a 24-inch length of flexible black insulated wire to terminal 8 of the switch and, after passing the end of the wire through a small hole in the box, solder this wire

to a black micro test clip.

     6.   Solder a 24-inch length of flexible red insulated wire to terminal 1 of the switch and, after passing the end of this wire through the same small hole used above, solder this wire to a red micro test clip. Also solder an insulated wire between terminal 1 of the switch and terminals 1 and 2 of the I.C.

     7.   Solder an insulated wire between terminal 5 of the switch and the black terminal of the push connector, and solder an insulated wire between terminal 2 of the switch and the red terminal of the push connector.

     8.   Solder an insulated wire from terminal 10 of the switch to terminal 1 of the pot. Now solder an insulated wire between terminal 3 of the pot., terminal 11 of the I.C. and terminal 4 of the switch. Also, solder the black wire of the battery connector to terminal 3 of the pot.

     9.   Solder an insulated wire between terminal 3 of the I.C. and terminal 2 of the pot.

     If you are converting a Citizen adapter box you would solder switch terminals 2 and 5 to the positive and negative plugs respectively, which fit into the meter jacks.

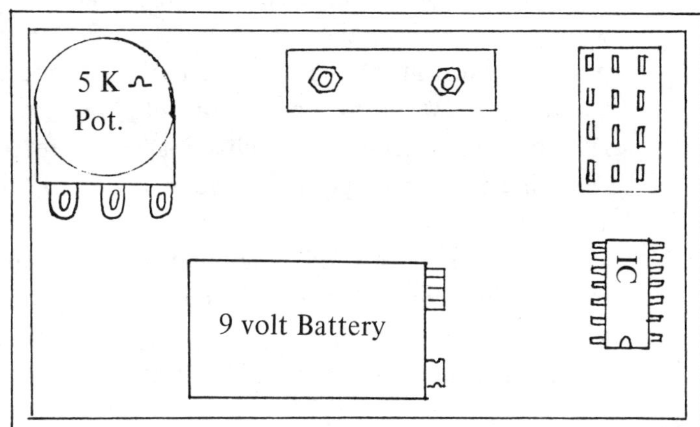

FIGURE 2

The Underside of the Lid of the Box

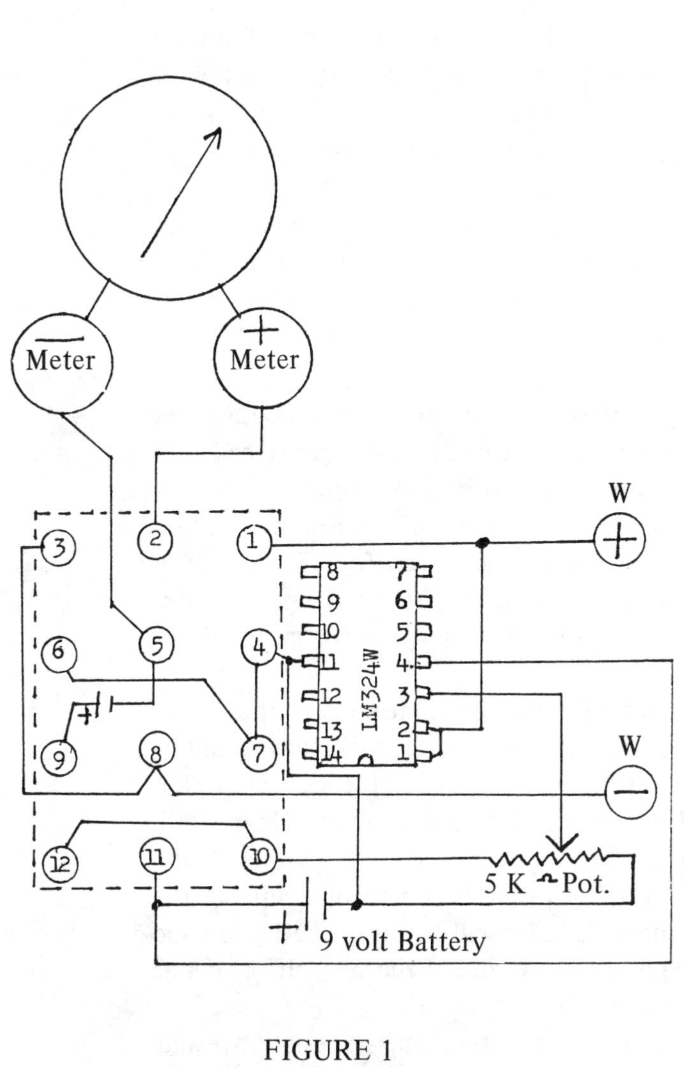

FIGURE 1

Wiring Diagram

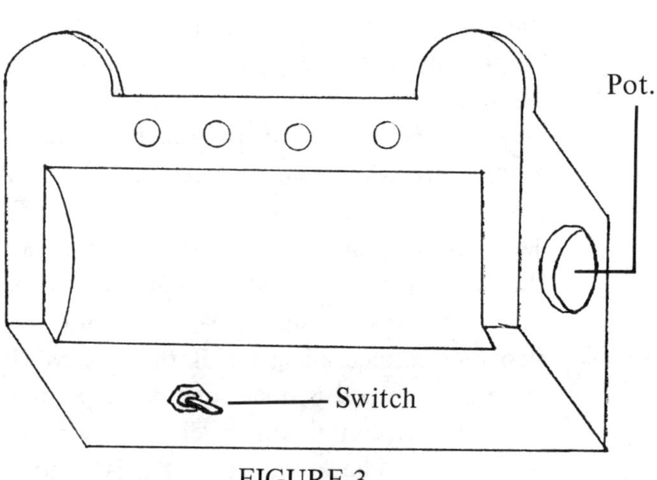

FIGURE 3

Citizen Attachment

Now, with some double-sided tape you can stick the I.C. and the battery to the inside of the box, taking care to see that they do not interfere with the pot., switch, and push connectors. All that is left now is to close the box and plug in the meter leads to the box. The micro test clips are used to connect the box to a watch you wish to test.

Now how do we use the meter and our new variable voltage attachment? When a customer brings in a watch for repair, the first test we should make is to check the cell voltage before we remove the cell from the watch. To do this we would set the variable voltage adjustment fully counterclockwise, set the adapter box switch to the left (or volts) position, and set our meter function switch to the lowest DC voltage scale above the expected cell voltage. We can now connect the red micro test clip to the positive side of the cell and the black micro test clip to a negative point on the watch. The meter will now register cell voltage.

To test the output of the circuit to the coil, you need only place the meter function switch to the lowest DC voltage setting and place the adapter switch in the left (or volts) position. Then, with sewing needles placed in the micro test clips, touch the needle points to the coil connection pads and watch the meter pointer swing alternately upscale and downscale with each pulse of the circuit output.

To power a watch with the variable voltage supply while monitoring the voltage, you would first remove the cell. Now connect the red micro test clip to the stem (or any positive point on the watch) and connect the black micro test clip to the negative cell contact. With the adapter box switch in the left (or volts) position and the meter function switch set to the lowest DC volt setting above the cell voltage, rotate the variable voltage pot. slowly clockwise while watching the meter pointer. You will be able to supply any desired voltage between zero and 7.5 volts.

Now is a good time to check for train freedom. While the watch is connected in this manner you simply lower the voltage slowly until the watch stops stepping; then raise the voltage slowly until the watch starts to step again. At this point, note the voltage indicated by the meter. A watch in good order should run at 80% of the normal cell voltage. (Example: 1.5V X .80 = 1.2 volts.)

Next, to measure current consumption, connect the red micro test clip to the stem (or any positive point on the watch). Connect the black micro test clip to the negative cell contact. Now, with the meter function switch set at the lowest DC volts position above the cell voltage and the adapter box switch in the left (or volts) position, watch the meter pointer as you rotate the variable voltage knob clockwise to 1.5 volts. Now switch the meter function switch to the microAmp position and change the adapter box switch to the right (or Amp) position. After a brief period of time, the meter pointer will settle down and indicate the current consumption.

Now to cover the recommended method of testing a coil for continuity without separating the coil from the circuit. To start, you rotate the variable voltage knob fully counterclockwise, set the adapter box switch to the right (or Amp) position, and set the meter function switch to DC microAmps. Then short the red and black micro test clips together. Now rotate the variable voltage knob slowly clockwise until the meter pointer reaches the right-hand side of the meter (full scale). Now without changing any setting, place the micro test clips on the two coil connection pads. If there is continuity in the coil there will be an indication on the meter. If the coil is open, the meter pointer will remain at the left end of the scale. If the coil is shorted, the pointer will travel all the way to the right side of the meter.

To check for a coil that is grounded, you will leave all settings the same as above and place one micro test clip on the pillar plate and the other micro test clip on one of the coil connection pads. At this point the meter pointer should not move. If the meter pointer does move to the right, this would indicate a grounded coil.

This should cover all the tests that can be performed on a quartz analog watch with a multimeter.

The only precaution to be taken with this adapter is to be sure that the switch is left in the center, or off, position when not in use. This will insure that the battery will not be drained.

Now!! Who said that watchmakers cannot be electronic technicians? And aren't you proud of yourself?

IN THE BEGINNING

By Gerald G. Jaeger, CMW, CEWS, FAWI

Give the watchmaker a series of wheels and pinions, a new type of automatic wind device, a different approach to driving a day or date ring on a calendar device, and he will study it in order to solve its intricacies. This is an old-hat problem for the watchmaker, because it is primarily mechanical, and it can be seen with the eye. This is what the watchmaker thrives on; he is trained in the art of mechanical problem solving. The watchmaker masters mechanical watch repair because he is able to apply mechanical concepts and principles to his trade.

However, these same concepts and principles do not apply to electronic watch repair. The difficulty in electronic watch repair is that electrons cannot be seen as they go about their work. As mechanical watchmakers, we were sure that if we ignored the electric watch long enough, it would fade away and never be heard from again. It is true that we cannot see the electron at work; however, the electronic watch has not faded away.

It is true that the first mass-produced and mass-marketed Hamilton 500 watch met with great consumer acceptance. The watchmaker found this watch hard to accept due to the difficulty in making the proper adjustment of the switching method used in the design of this watch. Many will recall that the contacts were fixed at the end of two extended wires attached to a bridge. The manufacturer apparently realized this was an area that could stand some improvement, so they made a design change. The new model became the 505, and was heralded as an improvement over the 500 model. As a favor to the watchmaker, the contacts were now on an index wheel and a roller table tab. Shortly after this, another favor was bestowed: both models were taken out of production.

The electric watch evolved into the electronic watch, and the electronic watch is now and will continue to be the predominant timepiece. Who can predict what form it will appear in? No one really knows, but we can be reasonably sure that the electron will be the charge that drives the display on both the contemporary watch and the watch of the future.

As watchmakers, we accept the fact that there are laws which apply to mechanical devices. It should not be too difficult for us to understand that electricity is also a science, with specific laws to which is must adhere. Our study of electricity will concern itself, in a most basic manner, with voltage, resistance, and current. When voltage, resistance, and current are understood, we will understand Ohm's Law, which is applicable to the everyday problems encountered in the repair of electronic timepieces.

It is not enough to know what voltage (volt), resistance (ohm), and current (amp) are; each must be seen. We can "see" volts, amps, and ohms when we learn to read the VOM (volt, ohm, milliamp meter). You will also see the VOM referred to as a multimeter. In this series of articles the meter will be referred to as a VOM.

There are three measurable quantities which, when combined, are electricity as it will be presented in this series of chapters. These measurable quantities are volt, amps, and ohms.

The VOM

The VOM is the instrument we will use to measure the three measurable quantities of electricity. We will approach the use of this meter in perhaps a manner not consistent with the norm, but which is consistent with the format of this primer.

The VOM is, in fact, three separate meters. In the beginning, it is most important that we learn to separate the three meters. To separate the three meters we will have to examine both the selector scale and the readout of the VOM. See Figures 1 and 2. To get an insight into this separation, we will examine both displays as a whole.

The selector scale in Figure 1 gives us options of selecting DC volts, DC amps, ohms, or AC volts. We will disregard the AC volts and treat the 3 to 6 position as though it were not on the selector scale. At this time, we will also disregard the 6 to 9 position. This area allows for the selection of DC amps (current). The 1 to 3 position, which will be discussed later, allows for the selection of ohms (resistance). This area is the ohmmeter area. The 3 to 6 area is alternating current, and even though AC really keeps the electron jumping about, we should not concern ourselves with AC at this time. The 6 to 9 position is another separate meter. It is an ammeter. It measures current flow which is measured in amperes. This area will also be discussed at a later time. Let us now consider the 9 to 12 position, which is the selector for the voltmeter.

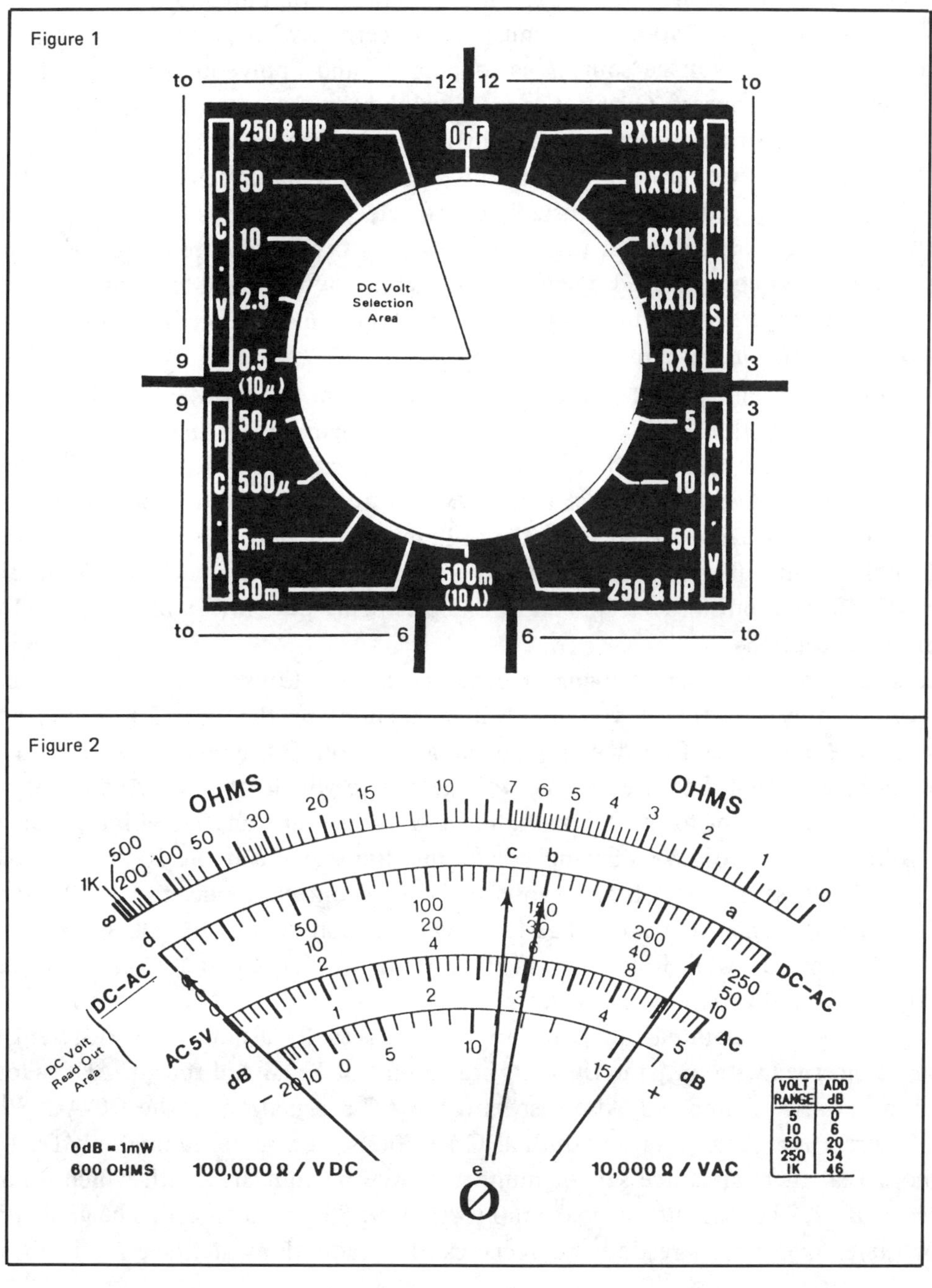

VOLT RANGE	ADD dB
5	0
10	6
50	20
250	34
1K	46

Voltage

Voltage, the moving force behind the electron, is comparable to the pressure in a water system. In a water system, voltage would be like pounds of pressure per square inch. Voltage is that comparable pressure which moves the electrons in an electrical circuit. Likewise, the electron flow could be compared to the water flow. This pressure which moves the electrons is measured on the VOM. At this stage of our study, we will address ourselves to the measurement of this pressure we call voltage.

All VOMs come equipped with two jacks or probes. One is black and the other is red. Note the two jack positions on the panel in Figure 3. One is marked +V-Ω-A, and one is marked −COM. In DC voltage, the red jack is the positive (+) and the black jack is the negative (−) or ground. The red jack will be inserted into the positive outlet, and the black jack into the negative outlet on the panel. For our purposes, we will consider the electron flow to be from negative to positive, so the pressure or voltage will be applied in that direction.

When measuring DC voltage we must be concerned with polarity (direction); therefore, we must examine all voltage sources as to polarity and apply our probes at the proper position of these voltage sources. On most current watch cells, the outer case is positive. The red probe would be applied to this outer case. The little tab or button at the top of the watch cell is negative. The black probe would be applied to this area. The C and D cells, which are commonly used in flashlights and other devices, are usually constructed in the opposite manner, with the little tab at the top being positive rather than negative. Most other cells are quite visibly marked. If the proper polarity is not observed when the probes initially make contact with the VOM, the pointer, or needle in Figure 2 will move in the wrong direction. To remedy this, simply reapply the probes, observing the proper polarity.

If you have a meter similar to the one being used in this series of articles, be sure that the upper right-hand position is set on +DC, as shown in Figure 3. At this time, do not be concerned with the other jack positions in Figure 3, since they apply to other meters, and will be referred to in due course. Please remember that VOMs are referred to as a group of different meters. At this time, we are concerned with the voltmeter, and only the voltmeter. I would be remiss if I didn't point out that your meter may vary somewhat from the one pictured and referred to in this article, but all concepts and principles being put forth will apply to all meters with few or any exceptions.

The selection area shown in Figure 1 contains the selections for ohms, amps, and volts. We are concerned only with the DC volt area, which encompasses the 9 to 12 position. Note that there are five different ranges from which to make a selection. If the selector were positioned at the 0.5 position, we could measure any DC voltage source which supplied a voltage from 0 to 0.5 volts. Moving the selector to the 2.5 position, we could measure any DC voltage source which supplied a voltage from 0 to 2.5 volts; and positioning the selector at the 10 position, we could measure any DC volts source from 0 to 10 volts. This same concept applies to the 50 and the 250 and UP positions. It stands to reason that for each selection area in the DC volts range there must be a readout area as well. Refer now to Figure 2. Note that there are six sets of readouts indicating figures from 0 to various maximums.

It was previously pointed out that the VOM is actually a number of different meters. Since we are concerned with only the DC volts portion of the VOM and the DC-AC readout area, disregard all other sets of numbers. Also disregard the AC designation of the DC-AC. When the selector in Figure 1 is positioned on any of the DC selections, we will be reading only DC voltages. You will note that there are three sets of numbers in the readout area with which we are concerned. See Figure 2. The readouts range from 0 to 10, 0 to 50, and 0 to 250. The graduations we will use for these three sets of readout numbers are the graduations that lie directly above the 0

Figure 3

to 250 set of numbers. This graduation will be used no matter which of the three sets of numbers are being used at a given time. The position on which the selector rests dictates the set of numbers which must be used. The number the selector rests on in Figure 1 will be found as the maximum number on one of the sets of readout numbers in Figure 2. If we would position the selector on 10 (Figure 1), then the 0 to 10 set of numbers would be used as the readout (Figure 2). This is the bottom row of the three rows of numbers in the DC readout area. This row of readout numbers must be used, since the meter is now set up to measure only from 0 to 10 volts. The other sets of numbers would not display a valid readout of the actual voltage. Were we to measure a voltage greater than 10 volts with the meter set at this range, we would very likely damage the meter.

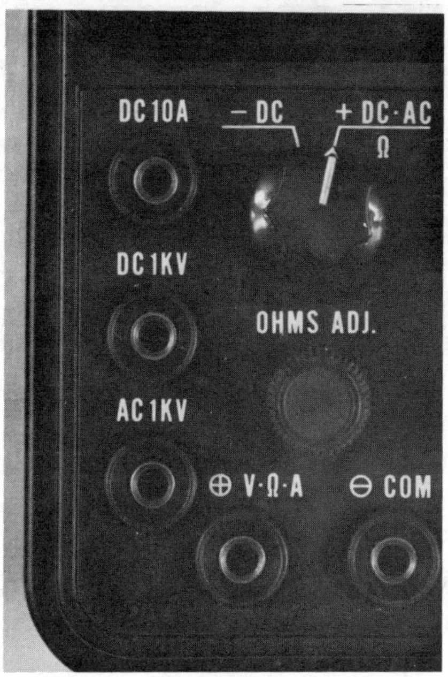

Another concept employed in VOM readouts is, perhaps, a bit difficult to readily understand. With a bit of practice and patience, it is not as difficult as it appears at first blush. Note that there are five selection areas in DC volts (Figure 1). Of these five selections, only three appear as the maximum voltage in our readout area (Figure 2). They are selections 10, 50, and 250 volts. Our three rows of readout numbers can handle these selection ranges, since one row of numbers is for 0 to 10 volts, another is for 0 to 50 volts, and the other is for 0 to 250 volts. Were we to measure a 9 volt battery, we would set the selector at 10 volts DC and use the 0 to 10 readout row of numbers. The pointer would register 9 volts. This is indicated by point "a" in Figure 2. In this case, we have completely disregarded the sets of 0 to 50 numbers and the 0 to 250 numbers and have used the 0 to 10 set of numbers and the graduation line directly in line with the DC-AC readout designation which is underlined in Figure 2. When we are in the 10 volt, the 50 volt, or the 250 volt selection ranges, it is a rather straightforward procedure, as the selections agree exactly with the readouts.

Let's assume we are measuring a 1.5 volt watch cell. The 10 volt selection range would handle this cell, but it would not be a good range to use, since the value of each increment on the readout line is 0.2 volts when the selector is set at 10 volts. We need to find a selection closer to the maximum voltage we expect to read. If we scan the selection ranges in Figure 1, we can find a 2.5 volt maximum voltage selection. But when we refer to the readout numbers in Figure 2, we do not find a 0 to 2.5 volt set of readout numbers. The three maximums in the readout are 10, 50, and 250, but there is a 2.5 volt maximum in the selection area of Figure 1. This indicates that there must be a 0 to 2.5 readout somewhere in the DC volt readout area. Here is where we must do some mental transposition. Suppose we look at the 0 to 250 set of numbers in Figure 2. Were we to drop the 0 from 250, we would have a 25. Now, if we place a decimal between the 2 and the 5, we would have a number 2.5 as a maximum on this set of numbers. This is exactly what the voltmeter has done when we rested the selector at 2.5 in the selector area. It has changed 250 to 2.5, and we must mentally do likewise. The readout numbers, as they appear in Figure 2, are 0, 50, 100, 150, 200, and 250. We must mentally transpose this line of numbers to read 0, 0.5, 1.0, 1.5, 2.0, and 2.5 volts. This selection and readout now become a maximum 2.5 voltmeter. Each increment on the readout line now represents 0.05 volts. We can now get an extremely accurate reading using this selection and this set of numbers.

Now, were we to measure a 1.5 volt cell, the indicator would rest as indicated at point "b" in Figure 2. Were we to measure a 1.35 volt cell the indicator would rest as indicated at point "c." The same concept applies to the 0.5 volt selection range. See Figure 1. When the selector rests on the 0.5 volt selection, this means we are in a meter area where the maximum, measurable voltage is 0.5 volts.

Referring now to the readout numbers we are using, we are again faced with a similar problem. Even though we have a selection to make the meter a 0 to 0.5 voltmeter, there is not a matching set of readout numbers, and we know there has to be readout numbers to match selection ranges. Once again, the transposition concept is applicable. We do have a 0 to 50 set of numbers on the readout. If we drop the 0 from 50, we have a 5. If we place a decimal point before the 5, we have a 0.5 as the maximum for this set of numbers. Again, let me point out that when we rested the selector on the 0.5 selection range, the meter already accomplished this transposition within the meter. Instead of this row of numbers reading as it actually does in Figure 2, we must now see it as 0, 0.1, 0.2, 0.3, 0.4, and 0.5 volts. Using this concept, we can now handle five different selection ranges with only three sets of numbers. You will see the necessity for this kind of number rearrangement when you consider there are 19 different selection ranges on the meter we are using. If we were to attempt to get 19 different sets of numbers on a readout, the size of the readout would be as large as your kitchen table.

Your meter, or the meter you may purchase, may not have the same selections or the same value sets of readout numbers as the meter we are using for this series of chapters. Do not be concerned. Upon close examination of your meter's selection area and readout area, you will find there will be a set of readout numbers to match your selection ranges. However, you may have to use the transposition method to accomplish this matching. We are using this meter only because its many ranges enable us to point out areas you will need to understand. If you are considering purchasing a meter, I would suggest purchasing one which is made specifically for use in watch repair. The basic meter concepts and uses discussed in this series will apply to all meters. I would discourage the purchase of a digital-type meter, unless you already have a good understanding of basic electricity and are already quite competent at meter reading.

You will not become an expert by reading a few articles. You will have to begin by measuring a few voltage sources. Step One is to adjust your meter pointer to 0 if it does not rest exactly on the 0s, on the left side of the scale (see Figure 2, point "d"). If your pointer needs adjustment, it can be accomplished by turning the screw in the lower center of the meter face (see Figure 2, point "e"). When measuring voltage sources, and the approximate value is known, set the selector at a range above but as close as possible to the expected voltage. If the expected voltage is not known, it is best to start at the highest range and work down. We could measure any voltage under 250 volts on the 250 volt range. Were we to measure a 9 volt battery at the 250 volt selection, it would barely move the pointer and the reading would be next to impossible to interpret. We want to have the meter set so we can obtain a reading in the upper one-half of the scale.

Read the instruction manual which came with your meter. Secure a few varied voltage sources—and "happy measuring"!

RESISTANCE AND THE OHMMETER

By Gerald G. Jaeger, CMW, CEWS, FAWI

Having completed our study of the volt and realizing that voltage does not move about in the wires of an electric circuit, we will undertake the study of resistance.

In the scheme of components in a simple electric circuit, the voltage is the pressure that moves the electrons in that circuit. The electrons, in their movement through the circuit, are the little gems that do the work. Were they allowed to traverse the circuit in an uncontrolled manner, these electrons would be useless and therefore must be regulated.

The circuit could be compared to a garden hose. If the hose were connected to a source with great pressure and if the water were turned on full force allowing the water to flow at an uncontrolled rate, we would probably do more damage to the area we intended to water than had we not watered it at all. If we were to put a controllable nozzle at the end of the hose, adjust it to a rate of flow applicable to the job specified, the result would be as intended. This is the same concept employed in the control of electrons in an electric circuit. We have a cell (battery) which supplies electrons and provides the force (voltage) to move electrons in the circuit. The electrons leave the cell and move along the conductive paths (wires) provided for their travel (flow). In order to control this flow of electrons, blockages or resistances are placed in their path. These resistances can be predetermined or fixed in order to allow only the required amount of electrons to flow in the circuit. As horologists we must be able to measure these resistances.

The VOM and Resistance

We have now established that resistance is the opposition offered to the flow of electrons in an electric circuit. It is measured in ohms and is measured by an ohmmeter. Having completed, for the time being, the measurement of voltage, we will place the voltmeter upon the shelf and bring down the ohmmeter, referring to the selection and readout areas for ohms.

The same concept applies to the selection area of resistance (ohms) as the readout area of resistance. The selection area of the VOM for resistance is the 12 to 3 position in a clockwise direction as shown in Figure 1. The readout area for resistance is the scale on the very top of Figure 2, parenthesis A.

A few things must be considered in order to understand ohm measurement. When measuring voltage the scale begins at 0, on the far left of the readout; the maximum voltage is read at the far right of the voltage readout. Noting the ohms readout we see that 0 ohms is found on the far right of the scale. This seeming contradiction of the meter readout will be clarified with further exploration of ohm measurement.

When measuring the resistance of any given component within a circuit, we are measuring the opposition to the flow of electrons offered by that component. In order to measure this offered opposition, we must attempt to move electrons through that component. This is exactly what we do with our selector set at any of the ohms selections. The VOM has a battery or two batteries housed within it. These batteries come into play only when we use the VOM as an ohmmeter. When the ohmmeter selector is positioned at any of the ranges, electrons are trying to flow from the black meter probe to the red meter probe. With the two meter probes lying next to the meter, free of contact with one another, you should have a meter reading of infinity. Note the position of pointer "a," Figure 2. Pointer "a" is at the far left of the ohms readout. Infinity is indicated by the symbol ∞. The number value of the resistance in ohms with the pointer at this position is so great that it is immeasurable. For our purposes we will consider this to be

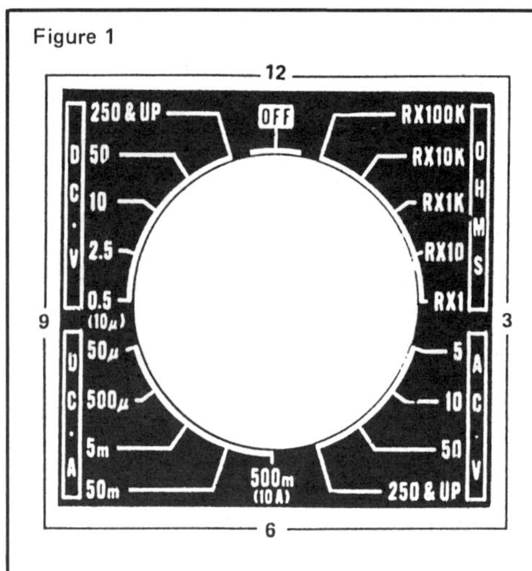

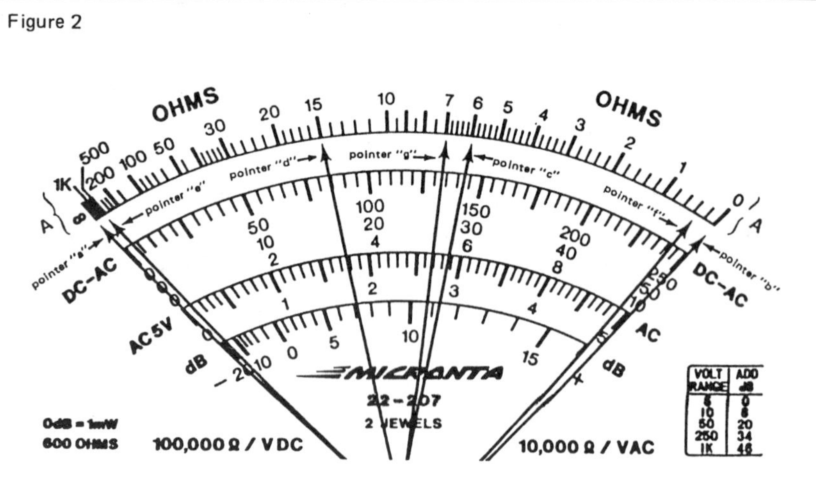

an open circuit. An open circuit is a circuit in which no electrons can flow because of a break in the conductive path which the electrons usually follow.

Our next task will be to have the two meter probes make contact with one another. This may be done by either holding them together with the thumb and forefinger of the left hand or connecting them with a jump wire. Set the selector at R x 1 in the ohms selection ranges. Find the ohms adjustment knob on your panel. The ohms adjustment knob is indicated by the arrow in Figure 3. The ohms adjustment knob may be on top of the panel, or it may be on the side of the meter. It is usually red and clearly marked. With the two meter probes making contact, adjust the ohms adjustment knob until the pointer on the meter points to 0 on the ohms read-out (pointer "b," Figure 2). If by turning the adjustment knob to a maximum in both directions the pointer will not "zero out," the meter will have to be opened and the batteries checked. If the batteries are good, check the battery contacts. If the battery(ies) is/are more than two years old, it is best to replace it/them with a new cell(s). Repeat this zeroing procedure at each ohms selection on the meter. Notice that each time you make a selection change, it may be necessary to rezero the needle. When measuring resistance and the necessity arises to change selections, it is imperative that we readjust the meter to 0. When the probes are making contact with each other and the pointer is at 0, this indicates the path for the flow of electrons is resistance free. The electrons are flowing from the negative terminal of the cell within the meter to the positive terminal of the cell, and the meter is measuring the resistance they encounter.

Continuity

We use the resistance measurements for two purposes: to determine the actual resistance offered by a component, such as a resistor or coil; and to measure continuity, which does not involve any actual numerical measurement of resistance. It is of prime importance to remember that any current other than that provided by the ohmmeter cannot be present in the circuit or component being checked or measured.

The measurement of continuity is employed as a check to see if a given path will allow electrons to flow freely. This check is used when we do not expect any resistance to the electron flow. We have seen that when the two meter probes are in direct contact with each other and the ohmmeter is set at any of the resistance scales or selections, the pointer registers 0 resistance (pointer "b," Figure 2). Were we to take a six-inch piece of wire with the two ends exposed and the portion from end to end hidden by insulation or wrapping, we could not visually see if this wire were continuous or broken. The ohmmeter could *see* this. By making contact with one probe of the meter and one end of the wire and making contact with the other meter probe and the other end of the wire, the meter would indicate if there were a continuous path or not. With the meter zeroed, the selector set at one of the ohms selections, and observing the meter readout, we too can *see* the condition of the conductive path. If the needle settles on 0 in the ohms readout (pointer "b," Figure 2), we can be assured this is an unbroken path allowing free electron flow. If the pointer settled at ∞ (infinity)—the position of pointer "a," Figure 2—this would indicate that a break or "open" was present in the wire. We cannot *see* if a solder joint is good. We cannot *see* if a hidden wire is complete from end to end. The meter can, and when properly used it can be an eye that can *see* the hidden.

Resistance (Ohms) Measurement

As in the measurement of voltage, the ohmmeter has a readout for every selection found in the selection area, even though there is only one set of numbers comprising the ohms readout (parenthesis A, Figure 2). Your meter may not have the identical selections of the meter being used for this primer, but the concept explained here is applicable with few exceptions. There are meters on the market that have 0 ohms on the far left of the readout scale and ∞ at the far right. With an understanding of resistance measurement, you should be able to apply these measurement concepts to this type of meter.

In reference to Figure 1, the 12 to 3 position will reveal five possible ohms selections. The selection we use will dictate the value of the readout for that specific selection. (We should, by now, realize that any further reference to the ohms readout refers to the ohms readout scale pictured at the very top of Figure 2, parenthesis A.) To measure a coil we must seek out the wire or conductive path entering the coil and the wire or conductive path leaving the coil. With the selector on R x 1 (resistance multiplied by one), the meter probes on the entering and leaving points of the coil wire, and with the meter already zeroed, we will begin to explore what the meter tells us. The selector set on R x 1 indicates that the meter readout should be read as the raw or actual numbers as they appear. The pointer resting at 6 (shown by pointer "c," Figure 2) indicates that there are 6 ohms of resistance present in the coil being measured. Were the pointer to rest somewhere between 6 and 7, the value of each increment would have to be calculated. With the selector at R x 1 the value of each increment would have to be 0.2 ohms.

Now measuring a different coil, with the selector set at R x 10, the pointer settles at 15 on the readout scale. This tells us the value of the coil is 15 times 10. The resistance within this coil is 15 ohms times 10 and has a resistance value of 150 ohms. Were the pointer to rest at one of the increments that lie between the 15 and 20, we would again have to make some mental calculation. Realizing that the selector being on R x 10 changes the value of the 15, the pointer rests from 15 ohms to 150 ohms. This concept applies to every number on the readout scale. With this in mind the 20 ohms now becomes 200 ohms. This then gives each increment between the 15 (actually now 150) and the 20 (actually now 200) a value of 10 ohms. This concept of readout value change with each selector change can cause confusion because of the mental multiplication required. I find it much easier to deal with the addition of 0s to the raw readout numbers as the selector indicates. When the selector is positioned at R x 1, the value of the readout scale is its actual raw value. In other words, a 6 is 6 ohms, 20 is 20 ohms, etc. When we move the selector to R x 10, the value of each readout number has one 0 added to it. A 6 becomes 60, a 20 becomes 200, etc. When the selector is set at R x 1K we add three 0s behind the raw number. A new factor to be considered is the value of K, which is 1,000; therefore, 7 would become 7,000 and 15 would become 15,000, etc.

We could go through all the selections of the ohms selector but that would require three more pages and another hour of reading. We can simply use the following mental transpositions rather than attempting to multiply. Following is a list of selector settings and the number of 0s to be added to the raw numbers of the readout.

R x 1:   Read out the raw values of the readout.
R x 10:   Add one zero to the raw numbers of the readout.
R x 100:   Add two zeros to the raw numbers of the readout.
R x 1K: Add three zeros to the raw value of the readout.
R x 10K: Add four zeros to the raw numbers of the readout.
R x 100K: Add five zeros to the raw numbers of the readout.

Keep in mind that as each readout value changes due to selection change, so too does the value of each increment between the numbers change. You will have to become adept at figuring this value change. Simply apply the concept as described in the earlier explanation of this increment change from a setting of R x 1 to R x 10.

When measuring resistance we want to get into the practice of measuring these values from the middle to upper (far right) regions of the scale. For example, we could measure a 500 ohm resistor or coil on the R x 1 selection. The pointer would rest on the 500 (pointer "e," Figure 2). We could also measure a 500 resistor or coil on the R x 1K selection. At this setting the pointer would rest at the position of pointer "f," Figure 2. This would be the better of the two choices since the readout is at the upper portion of the readout scale. Most meters have a R x 100 selector, and this would be the ideal selection as it would put the pointer at the raw 5. With the selector at R x 100, the raw 5 on the readout would translate to 500. It will be noticed that many of the resistance measurements can be made in more than one selection area. Always use the area that will result in the pointer settling at the mid to upper portion of the readout scale. This practice must be followed because the meter is more accurate in this readout area, and the incremental value between numbers allows for a more concise determination.

Let's look at one more resistance readout. Assume we were measuring a resistor or coil with a resistance value of 7,000 ohms. We would set the selector on R x 1K. The selector set at R x 1K would, in effect, append three 0s to the raw 7 on the readout, making the 7 a 7,000. The pointer would rest at the position of pointer "g" in Figure 2. Should it rest a bit under or over this position, we would have to ascertain the value of the increments between the 6 and 7, or 7 and 8. Notice there are five increments between the 6 and 7 on the readout. Knowing that the *actual* value of the raw number 6 on the readout is 6,000 ohms and the *actual* value of the raw number 7 is 7,000 ohms, dictates that each increment between the 6 and 7 on the readout scale has a value of 200 ohms. Whenever the pointer falls between a raw number or an incremental marker, an educated guess will have to be made as the exact value being registered.

When reading resistances it is well to keep in mind there is an acceptable error margin. Suppose the resistor had a 10% error or the coil had an allowable 10% error, and suppose the meter had a 5% error and we erred 5% in reading the meter. This adds up to a 20% error, should all errors be made in the same direction.

Now suppose we were measuring a coil that a technical bulletin reported as having a value of 3,000 ohms. If we measure this coil and come up with a reading from 2,700 ohms on the low side to 3,300 ohms on the high side, we would consider this to be a very acceptable reading as it is well within 20%. In fact, technical guides allow for this amount of error and state the coil may vary even more than the 20%, making the watch well within the acceptable operating coil resistance ranges. A technical bulletin reports that the coil value is acceptable when it measures from 3,400 ohms to 4,300 ohms. By pointing out that there are allowable variations in ohms

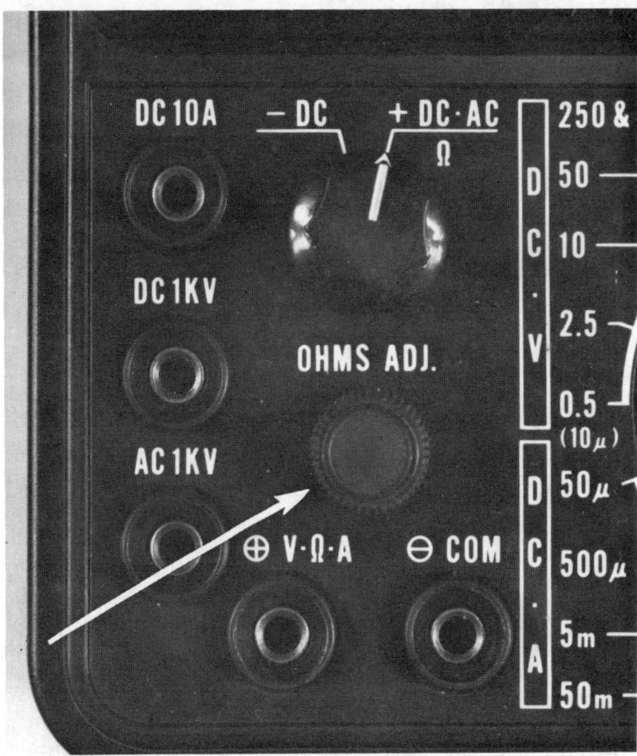

measurements by no means implies that poor measuring practices are acceptable. A complete understanding of how a low or high reading coil affects the operation of a step motor watch will be a great aid in troubleshooting for difficult-to-find problems you may encounter when undertaking the repair of these watches.

You have probably noted that in these chapters I do not show a meter specifically for horological application. There are a few reasons for this. My meter study and its application to horology began before some of the presently available meters were on the market. I continue to use this meter since its wide range offers some teaching opportunities others do not. I am not in a position to recommend any specific meter, but I would suggest that meters which are available from watch material supply houses or watch manufacturers be examined. They are more applicable to the volts, ohms, and amps ranges encountered in watch diagnosis and repair. They also have or are readily adaptable to some add-on attachments which are being introduced within the industry.

AMPERAGE AND THE AMMETER

By Gerald G. Jaeger, CMW, CEWS, FAWI

Having completed two of the three measurable quantities that make up the phenomenon referred to as "electricity," we will undertake the study of amperage. Electrons flowing in a circuit are what constitute the current within that circuit. This flow of electrons, being one of the measurable quantities, is measured in increments called amps.

The VOM and Amperage

The selection area of the VOM has a separate area for the measurement of amps. It is the 6 to 9 o'clock position as shown in Figure 1. There is, at this point, a seeming contradiction. The selection area is labeled DC A (Amps), but there is only one amperage selection, and it is in parentheses (10A). It is positioned at the 6 o'clock location, situated directly below the 500m selection. There are seven different selections which could be made in the amperage selection area. They are 10μ, 50μ, 500μ, 5m, 500m, and 10A. This would surely leave us in a quandry were we to make an attempt at measuring current, prior to some exploration into the values of current as indicated by the μ, the m, and the A.

All electron flow (current) is measured in amperes or percentages thereof. These "percentages thereof" are milliamps (mA) and microamps (μA). An understanding of amps, milliamps, microamps, and their value relationship to each other is absolutely necessary in order to successfully measure current flow with a VOM. This would be accomplished in a rather simple manner if we were familiar with the metric system. We are not, so let's try another approach.

In order to explain the value difference between amps (A), milliamps (mA), and microamps (μA), we will have to involve ourselves with considerable repetition. In fact, an ampere is a large volume of current flow; a milliamp is 1/1000th or 0.001 of an ampere; and a microamp is 1/1,000,000th or 0.000001 of an ampere. The VOM is constructed so that it does not measure amps in fractions or decimal equivalents as such. Actually it does; however, the current values are not read as such. The fractional or decimal equivalents of amperes are read in milliamps (mA) or microamps (μA). It would be impossible to relate exactly how much current flow constitutes 1/1000th or 0.001 of an ampere. We can relate more easily to another term, the milliamp.

The milliamp is 1/1000th of an ampere fractionally and 0.001 of an ampere decimally. It would obviously take 1000 milliamps (mA) to equal one amp of current flow. In horology, current flow in the full amp increments will never be encountered. However, current flow in increments of milliamps will be encountered. The meter is equipped to handle these increments. Milliamps are still very large amounts of current flow when related to horological application. The only time this great a current flow is encountered in today's timepieces is when a light or alarm is activated. The noisemakers that accompany some of the game watches now on the market would have to be considered as current consumers in the mA ranges. The new radio watch being marketed will also require these large amounts of current. I like to see these developments. They may not require great amounts of electronic repair, but there are a lot of pushers that will require service, and millions of cells will have to be replaced in order to keep these boys and girls who wear them happy. (Like they say, one difference between men and boys is the cost of their toys.) Watch cell installation is a quite reasonable profit item and an excellent store traffic builder.

In Figure 1, notice there are three possible milliampere selection ranges. They are indicated by the 5m, the 50m, and the 500m positions found in the DC amperage selection area.

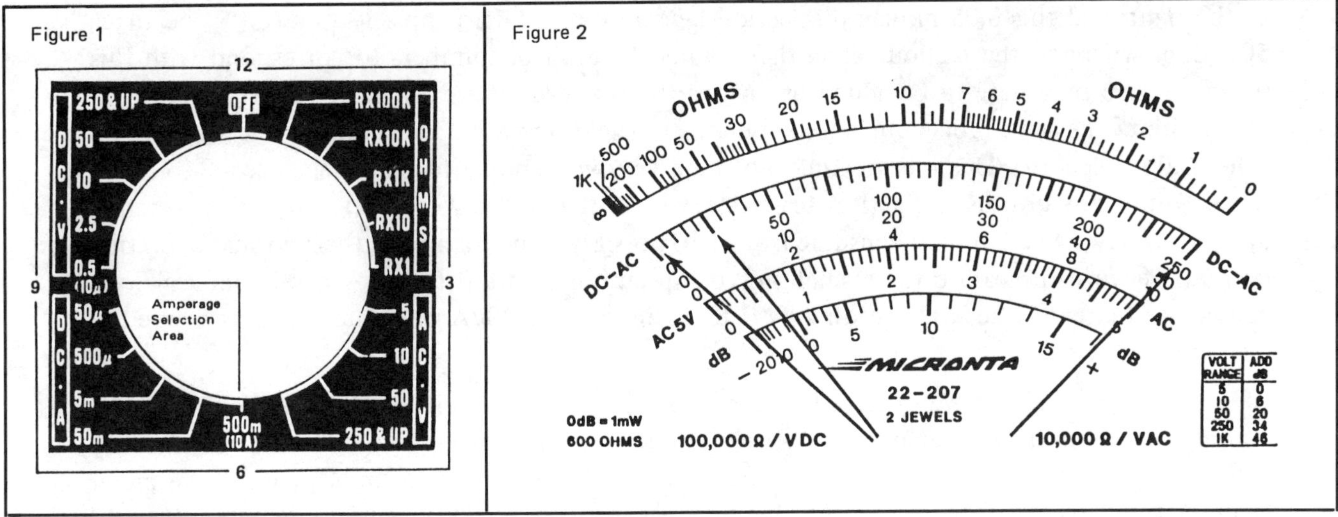

These choices indicate we can measure milliamps in the 0 to 5 maximum, the 0 to 50 maximum, and the 0 to 500 maximum ranges. The selection possibilities, as indicated by a study of the selection area in amperage, show there are actually one amp meter (A), three milliamp meters (mA), and three microamp meters (μA) all combined in the ampere portion of the VOM.

With reference to the readout area in Figure 2, please note there is not a readout area specified to be used for amperage measurements. When measuring amperage on most meters, this one included, the DC-AC volts area will be used. This area is noted on Figure 2. This is a readout area we are familiar with from the study of voltage. As in the voltage readout study, whenever a selection is available in the milliamp selection area, there must be a corresponding value range in the readout area. Using the transposition method, the 0 to 50 maximum set of numbers (middle row) will readily handle the three milliamp selections found in Figure 1. The transposition task is considerably easier here, because only one set of numbers will be used. If the selector is set on the 0 to 50m selection, the raw numbers in this middle row will be used. If the selector is resting on the 5m selection, the middle row will be treated as being a 0 to 5 milliamp maximum readout. Simply drop the 0 from each raw number. This changes the incremental value between each number and must be taken into account. If the selector should rest on the 500m selection, a 0 would be added to each of the raw numbers of the readout. The milliamp meter is now a 0 to 500 maximum milliamp meter.

The microamp is 1/1,000,000th of an ampere fractionally and 0.000001 of an ampere decimally. Notice that the prefix "micro" means millionth of. The prefix "milli" means thousandth of. A microamp is a very small unit of electron flow; actually one millionth of an ampere. It is the smallest unit of electron flow the meter is engineered to handle.

The amperage selection area of the meter being used in this primer has three microamp (μA) selection areas. They are 0 to 10μA, 0 to 50μA, and 0 to 500μA. These three selection areas can be easily matched in the readout area. See Figure 2. Both the 0 to 10 and the 0 to 50 can be found as raw numbers on the readout in the amp readout area. Should the selector rest on the 500μ selection, one 0 would be added to each raw number on the 0 to 50 row of numbers in the readout area. This readout set of numbers now becomes a 0 to 500 microamp readout and corresponds with the 0 to 500 selection.

There will be many VOMs that do not have a microamp (μA) selection. If the meter you are using does not, then examine the milliamp selections (mA). If it has a 0.05m or 0.06m selection, this selection area can be used for microamps (μA). There are 1000 microamps in a milliamp. If the 0.05m (0.05 milliamp) is used, the decimal can be mentally moved three places to the right, and this 0.05 milliamp selection becomes a 50 microamp selection (0.05 becomes 50). Somewhere in the readout area, there would be a set of numbers to correspond with this selection. A 0 may have to be added or dropped in the readout area to match the selection with the readout. The same procedure would apply if the meter has a 0.03m or 0.06m selection or any other 0.0m selection. This meter would not be a very good meter for our use, since a 0 to 50μA range would not give us a practical scale to read out the very low current consuming watches. Many of today's watches only consume from 1 to possibly 5 microamps as their normal operating current. If we can measure these small microamp currents on a 0 to 10μA maximum scale, we can get an extremely accurate reading. A 0 to 12μA or 0 to 20μA maximum will also serve the purpose quite well.

The VOMs I have observed that were manufactured specifically for horological use by watch companies have a separate DC amp readout scale. This readout corresponds with a very low 0 to maximum microamp selection position. The one I am most familiar with is the meter manufactured by the Citizen Watch Company. It has a 0 to 12μA selection area and a 0 to 12 maximum μA readout area. The Seiko meter is constructed in a similar fashion and also has this very desirable low microamp readout feature. Both meters have additional microamp and milliamp selections. They are read in the DC volts readout area.

The Seiko and Citizen meters also have another desirable feature. They color code their selection area with their readout area. The volts, amps, and ohm numbers are each a different color in the selection area and these colors correspond with the color of the numbers in the readout area. This scheme is helpful to the beginning electronic horologist but is not in itself a worthy factor in warranting the replacement of your present meter if it is satisfactory to you. As previously stated, if I did not have a meter and was contemplating the purchase of one, I would choose the Citizen or the Seiko type meter. Other meters are entirely satisfactory, although some attachments may have to be built to make them readily adaptable to the measurement of current. The Accutron meter by Bulova is also very adaptable to current measurement, and its adaption will be covered later.

Understanding volts, amps, ohms, and their measurement and applying this understanding to any and all meters is extremely important. I have found that the most confusing aspect of current measurement is understanding the difference of values (amounts of current) between the amp, the milliamp, and the microamp. One amp of current is equal to 1000 milliamps and is also equal to 1,000,000 microamps. By this same relative value, 1 milliamp equals 1,000 microamps (1mA = 1,000μA). Another approach would be to consider 1 amp. An amp being equal to 1,000 milliamps, consider 1.0 amp, and move the decimal point three places to the right. The 1.0 amp is now in increments of milliamps and shows that by moving the decimal three places to the right, amps are changed to milliamps. To change amps to microamps, move the decimal six places to the right. By the same number value, 1,000,000 microamps can be converted to milliamps by moving the decimal three places to the left (1,000 000. μA becomes 1,000. mA). Moving the decimal six places to the left will change microamps to amps (1,000,000. μA becomes 1A).

Moving the Decimal to Change A to mA to μA:

Moving the decimal three places to the right, 0.000 005A, changes A to mA (0.005 mA).
Moving the decimal three more places to the right, 0.005 mA, changes mA to μA (5.0μA).

Conversely:

Moving the decimal three places to the left, ⤳5.0μA, changes μA to mA (0.005mA).
Moving the decimal three more places to the left, ⤳.005mA, changes the mA to A
(0.000005A).

Realizing that 500μA is the same amount of current as 0.5mA, notice that either the 5m (0 to 5m) selection or the 500μ (0 to 500μ) selection could be used. Were we measuring a device which we knew was rated to consume 500μA, we could use either the 500μ selection or the 5m selection (see Figure 1). Let's explore which would be the wiser of the two choices.

Should the selector rest on the 500μ range in the selection area, this would dictate that the middle row of numbers in the readout area would be used. The selection of 500μ dictates that the readout area be considered a 0 to 500 microamp readout. Taking a measurement, the pointer would rest at position "a" in Figure 2. This would indicate a consumption of 500 microamps. The transposition has changed this 50 to 500.

Should we choose to measure this current flow in the mA rather than μA, the selector would rest at the 5m. This converts the ammeter to a 0 to 5 milliamp maximum meter. Taking a measurement, the pointer would rest at position "b" in Figure 2. This would indicate consumption of 0.5m (0.5 milliamps). We are using the same middle row of numbers, but the selection of 5m has dictated that this middle row of numbers (0 to 50) now becomes a 0 to 5mA maximum readout scale.

Even though 0.5 mA and 500μA are the same quantity of electron flow, one of the selections is a better one to use in the measurement than the other. Let's examine why. Using the 0 to 500μA scale, the increments between any two of the full numbers have a value of 10μA. Using the 0 to 5mA selection, the increments between any two of the full numbers on the readout scale have a value of 0.1mA.

Should the needle not fall exactly on an increment or a full number, it is easier to accurately calculate a part of 10 microamps than to calculate a part of the milliamps, since the milliamps are already only a portion of a whole number. This current flow could also be measured on the 50m selection. The needle would fall at position "c" in Figure 2. You can see how difficult it would be to make an accurate determination at this setting.

The measurement of current consumption is not difficult but requires a different technique than employed in voltage and resistance measurement. It is imperative that the horologist understands NOW the behavior and function of current flow and its measurement. Five years hence, the technical data available to the bench watchmaker will be beyond his comprehension should he neglect this basic study at the present.

CURRENT CONSUMPTION
AND ITS MEASUREMENT

By Gerald G. Jaeger, CMW, CEWS, FAWI

The measurement of electron flow is a straightforward procedure. The measurement of voltage and resistance is accomplished by the meter being parallel with the device being measured. Current is measured with the meter is series with the circuit being measured.

Figure 1 provides insight as to how this is accomplished. The load or watch is represented by letter "a," the cell by "b," and a 470 microfarad, 16-volt capacitor by letter "d." If we consider the electron flow in Figure 1, we will note the flow is from the negative ($-$) terminal of the cell (b) to the negative side of the capacitor (d) which becomes charged to a voltage equal to that of the cell. The capacitor's negative plate, holding all the negative charge possible, will then allow electrons to flow from the cell through the negative ($-$) probe of the meter, and through the meter, exiting through the positive probe (+) of the meter. The electron flow will continue to the negative ($-$) cell strap of the watch and exit the watch through the positive (+) cell strap, returning to the cell's positive (+) terminal. The connections all being previously completed as indicated in Figure 1, the ammeter is instantaneously recording the actual current flow, or as we refer to it, the consumption of the watch. The ammeter selector would be resting on the lowest μA (microamp) selection, while reading the consumption on the appropriate readout scale.

You will note the capacitor is connected parallel to the meter. The use of a capacitor is necessary when measuring the current consumption of a step motor watch. The duration of current flow which energizes the coil and drives the step motor is in the vicinity of 10 milliseconds. Some of the newer step motor coils are energized for a duration of under 4 milliseconds per step. Without the capacitor in the circuit, each energizing of the coil would cause the meter pointer to jump forward and settle back to 0 between pulses. The pointer would not reach a maximum reading of the actual current flow because the duration of flow is too short a period for the meter to respond. The capacitor will level out the action of the meter pointer, and it will slowly raise to the actual current consumption reading, settling there for an easily read indication of actual current flow. Some watches which pulse only once a minute or once each half minute may cause some difficulty in attaining an accurate measurement. (We will confront that problem when it becomes more prevalent.) Inserting another capacitor to be switched in when needed may be the answer.

I have been informed that some of our field watchmakers have experienced problems reading solid-state (LED and LCD) watch consumption when using the attachments containing capacitors. I have not found that leaving the capacitor in the circuit on solid-state watches to cause any particular problem. Consumption readings on solid-state watches do not require the capacitor. There are no surges of current in LED and LCD watches that are necessary in the step motor watches. If you are experiencing this problem, it would be a simple matter to attach a switch to any of the attachments which contain the capacitor. Simply switch in the capacitor when measuring step motor watches and switch it out for the solid-state watches.

Many watches, because of the higher starting current required, will cause the meter pointer to peg at the far right of the readout at first contact of the probes. This can be handled in a number of ways. One method is to leave the probes on the watch. Quite often the watch will start functioning, and the meter pointer will slowly settle back to the proper reading. Another method is to remove the probes from the watch. As the capacitor discharges, the pointer will slowly settle back to 0. When the pointer has returned to 0, reapply the probes.

If both these methods fail, you can attempt to get the watch started on a milliamp selection of the meter. To accomplish this, put your meter selector on one of the lower milliamp selections. Place the probes on the watch and after it begins to function, switch the meter to the proper microamp range. Do not remove the probes from the watch when changing the selector. One of these methods will generally work. Any time you have trouble with the meter pegging, you should first check the polarity of the meter probes relative to their point of contact in the watch. Continued pegging will occur if proper polarity is not observed.

Current measurement requires some method of energizing the movement and still being able to insert the meter in series with the circuit being measured. This can be accomplished in a number of ways. Figure 2 shows perhaps the simplest method. The meter shown has a special adapter attached to it which provides both a power supply and a capacitor housed within the adapter. By placing the appropriate cell in the clip provided, resting the selector on the correct μA selection and applying the probes properly, the watch will begin to function. The actual current consumption will be displayed on the readout. I want to stress again that in applying the probes to the watch, we must be constantly aware of polarity. The positive meter probe must be applied where the positive terminal of the cell would make contact in the watch. This also applies to the negative meter probe. (Those who have the excellent Citizen meter and attachments can refer to the instruction manual for other applications.) A milliamp range sufficient for measurement of night light bulbs, alarms, and other higher current consuming features is most desirable.

The Accutron test set, which many watchmakers own, is an excellent ammeter. Its capability in current measurement is very good. The test set pictured in Figure 3 shows its application. A cell must be inserted in the cell well of the test set. A yellow lead and a blue lead must be attached to the pictured adapter (No. 9920/6603). The adapter is no more than a capacitor and is so constructed that when properly connected, the polarities will be correct. With the selector set at "Read Microamps" and the red and black leads from the adapter properly applied to the watch, a current readout in μA will be present. Disconnection of the adapter will enable you to read LCD solid-state watches directly from the test set. Simply adapt the blue and yellow leads from the test set so they can be attached to the positive and negative cell leads of the watch. Measurement of LED quiescent current is possible. LED display current cannot be measured with this ammeter, because it does not have milliamp capabilities. These milliamp cur-

Figure 1

Figure 2

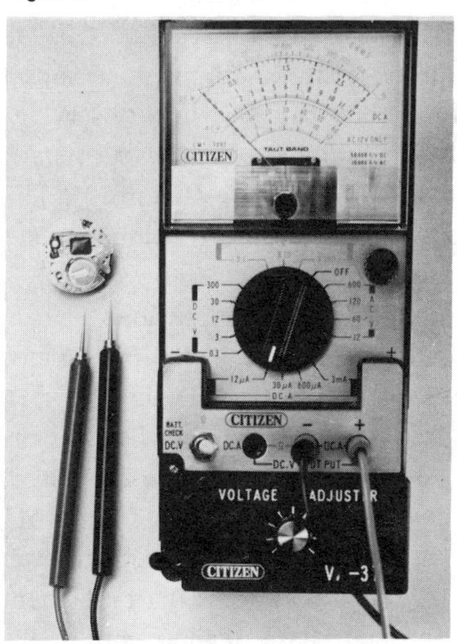

Figure 3

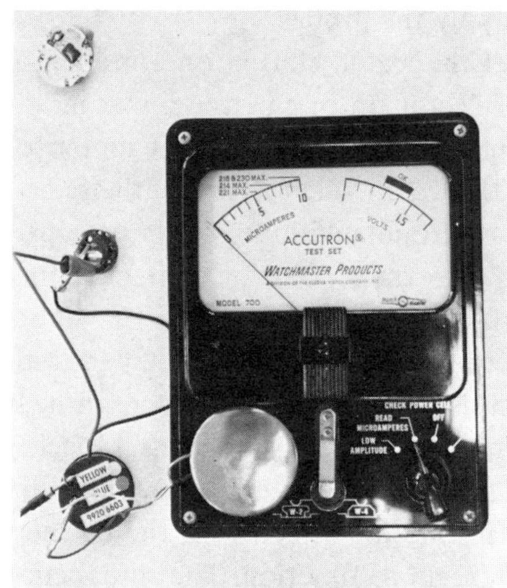

Figure 4

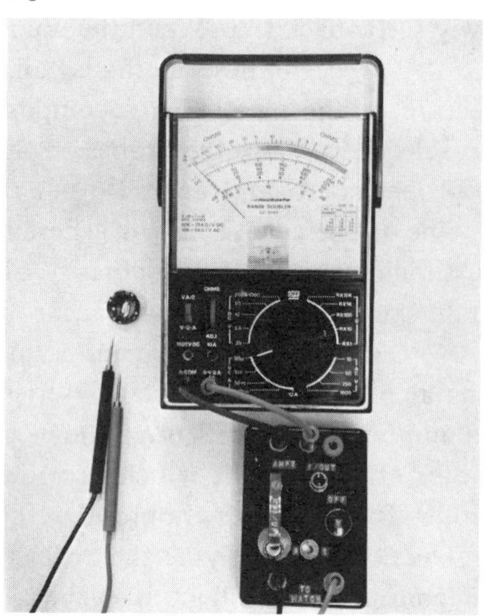

rents do not require a capacitor so they can be measured on a VOM. Most watchmakers who have an Accutron test set also have a VOM.

Many watchmakers are already owners of a VOM which is adequate for the measuring of voltage and resistance. This type of meter will need some sort of adapter for the measurement of current. Figure 4 shows such an adapter. This adapter has the feature of a current supply in the form of a receptacle to hold the proper watch cell. With proper connection, the adapter functions similarly to the diagram in Figure 1. Another feature is a short-out button for the capacitor. Earlier in this chapter we discussed a tendency for the meter to peg because of a higher starting current requirement. When the adapter probes are first applied to a watch, the capacitor charges to its maximum, then current begins to flow through the meter to the watch. Pressing this short-out button will discharge the capacitor and the pegged pointer will fall back to the 0 immediately. This avoids the unnecessary delay of waiting for the capacitor to discharge and the pointer to return to 0.

Some watch manufacturers, in their technical bulletins, call for a low-operating voltage test. A third feature incorporated into the attachment (see Figure 4) is a variable voltage supply. There are voltages below the 1.5V normal operating voltage at which the watch should function properly. This variable voltage supply will enable us to supply 1.5 volts to the watch and slowly reduce the voltage while observing at which voltage the watch ceases to function. Technical bulletins will state the lower operating voltages which are acceptable. The Citizen meter also has this desirable feature (Figure 2). This variable voltage supply is able to supply up to three volts and with some modification could be made to supply any voltage that may be required in the foreseeable future.

This complete adapter can be built for about $12.00. Those who built the earlier adapter, which appeared in the *Horological Times*, will be able to modify theirs at a lesser cost. Complete instructions explaining how to build and the components needed will either appear in a future issue of *Horological Times* or an instructional diagram will be available from AWI Central.

There are other practical approaches to the measurement of current consumption. Understanding the characteristics of electron flow is most desirable when attempting to troubleshoot for problems in electronic watches. The Seiko Time Corporation has two excellent technical guides which are part of their technical library. I have been informed they are available without purchase of the complete library. They are the *General Instruction, Seiko Analog Quartz* and *General Instruction, Seiko Digital Quartz*. They are precisely what the names imply: GENERAL INSTRUCTIONS to the repair of analog and digital quartz watches. These technical guides explain and diagram techniques beyond those I have outlined in this article. These guides give instruction of a general nature and the information is applicable to most watches regardless of manufacturer. Should you encounter a testing procedure that applies only to specific watches, you wil have to obtain the individual technical guide for that specific watch. General procedures are well covered in the two Seiko guides.

SIMPLE ADAPTER FOR MEASURING
CURRENT CONSUMPTION

By Edward F. Rice, Certified Electronics Technician
and
Gerald G. Jaeger, CMW, CEWS, FAWI

While it is not always necessary to measure consumption when diagnosing a battery-operated watch, certain failures can be easily isolated this way. The trouble is that the manufacturer's instructions for current measurement are often sketchy or nonexistent, and they may require that you use a special meter sold by the manufacturer.

If you have tried to measure consumption using your ordinary volt-ohm-amp meter, you found that on some balance wheel types you can't even get the watch to run when the meter is in the circuit. This is because the meter itself requires some power to move the needle and there just isn't enough to run both the watch and the meter. On other types of watches the needle jumps around so much that you can't read a value accurately off the scale.

The simple little adapter described here and shown in Figure 1 has two short leads extending from it to be plugged into your meter in place of the regular probes. Then, using the other test leads extending from the box, you will be able to measure current consumption in any watch. The cost of parts is about $6.00 and it will take about one-half hour to construct the adapter. Here's how to do it.

Figure 2 shows a top view of the plastic box with the dimensions we used for spacing the holes to match our meter. You may need to space the holes differently for your meter. The diameter of the holes depends on size of wire used.

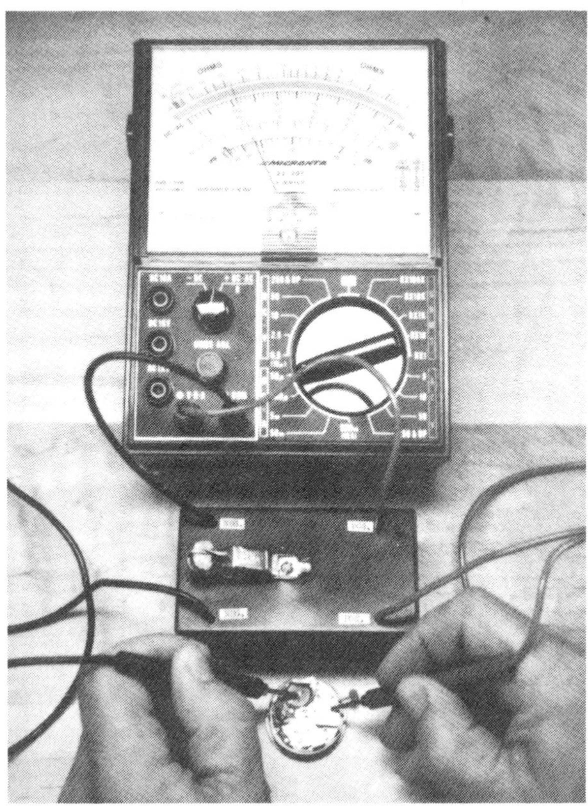

Figure 1

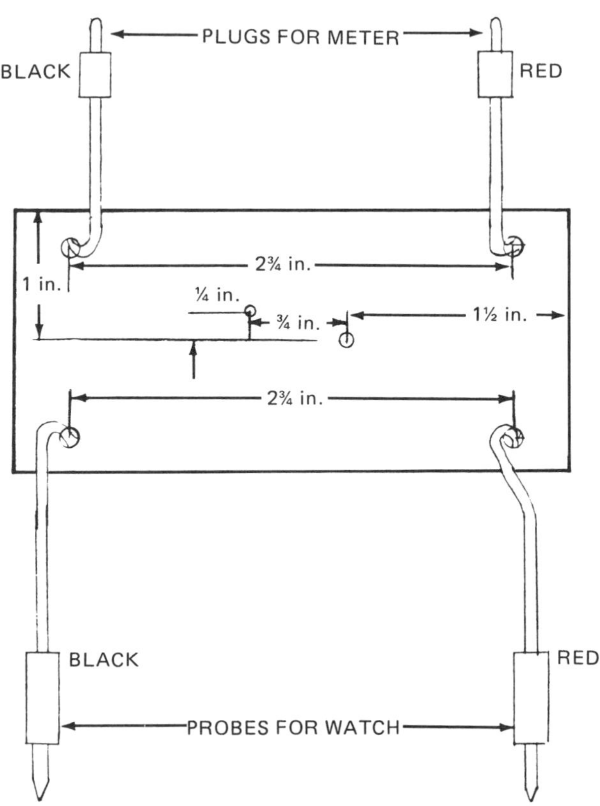

Figure 2. Top view.

The best way to provide the wires is to buy a set of ready-made test leads with the right kind of plugs for your meter. The pointed tips are never sharp enough for use in a watch so you will have to sharpen them on a grinder or bench lathe, or with a file. Cut each lead about 8 inches from the end that goes into the meter. This gives four pieces of wire. The two shorter ones will go through the holes at the top of Figure 2 and the longer ones will enter the box through the lower set of holes. We put the black wires on the left side to match the location of the negative connection to our meter. A little cement on the inside secures the wires nicely.

One of the two holes near the center is for mounting the battery clip and the other one is for a piece of wire. When the clip is mounted on top of the box with a machine screw and a nut, a wire from one of the jaws must be fed through the remaining hole to the inside, and another wire is secured to the nut holding the mounting screw. Since the clip will be mounted with only one screw, a little epoxy under it will keep it from twisting to the side when in use.

But before mounting the clip, it must be modified as follows to insulate the jaws from each other (see Figure 3). The type of clip specified in the parts list is the kind where the jaws open vertically, moving apart at 180 degrees, so they will bear perpendicularly on the battery surfaces. The "alligator" types won't work because the jaws open to an acute angle, like the jaws of an alligator, and therefore do not hold the battery securely.

Insulating one jaw from the other is a simple matter. First, remove the side and corner teeth from one jaw, leaving only two front teeth. This will be the lower jaw when the clip is mounted and it will contact the smaller negative terminal of the battery (for most batteries). Next bend all the teeth outward on the other jaw so the edge of this jaw is flat. Now insulate this edge by cementing to it a small piece of insulating material such as a rubber or fiber washer about the same diameter as a watch battery. Or, if you have handy some of that silicone rubber sealer that comes in a tube, a blob of this can be shaped to do the job.

The last step in the preparation of the clip is to cement on top of the insulator a brass washer large enough to make a contact for the battery when it is held between the jaws. Before cementing it, solder a few inches of thin insulated wire to the washer. This part of the clip will be the positive battery connection. The other side of the clip will be on the bottom contacting the negative side of the battery.

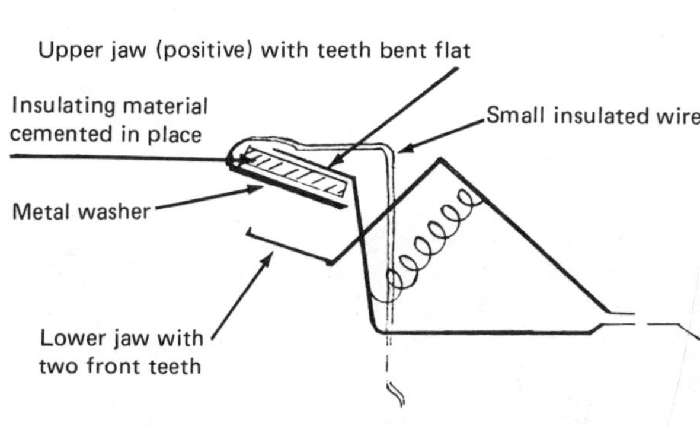

Figure 3. Clip modification.

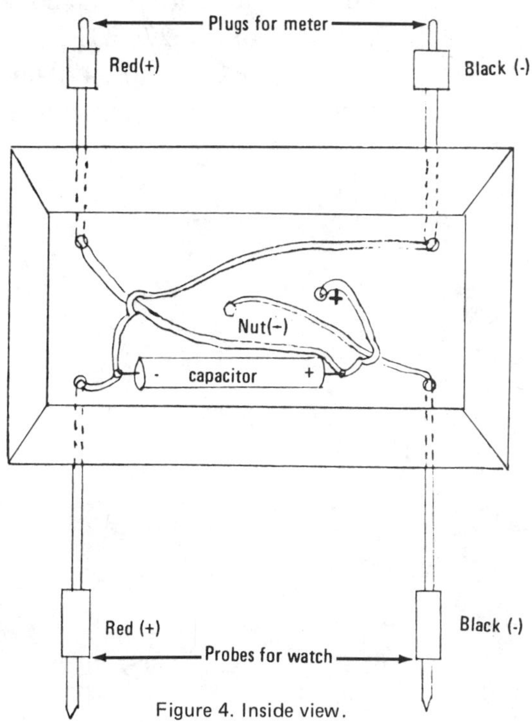

Figure 4. Inside view.

Alter all the parts are mounted they can be connected with short pieces of wire as shown in Figure 4. Be sure to note the polarity of the capacitor leads. Sometimes the leads are marked with plus and minus signs and sometimes only one lead is marked. The adapter will not work if the capacitor is wired backwards. Also, be careful about connecting wires which come through the box from the top so you don't mistake a positive for a negative one. Figure 4 shows the box with the red leads on the left side, but this is a bottom view. This arrangement matches our meter connections. You may decide to reverse these parts if your meter connections are the opposite way.

The principle on which the adapter works is not complicated. It depends entirely on the effect of the capacitor which is connected directly across the meter movement. When the probes are connected to a watch, pulsating current flows from the battery, through the watch, and also through the meter movement. The meter movement is simply a small coil mounted between permanent magnets so that it will turn when current flows through it. The indicator needle is attached to the coil. If it were not for the capacitor the coil would respond to the quick pulses of current drawn by the watch and it would be impossible to take a reading.

The capacitor acts like a storage tank. When electrons are flowing during a pulse the plates of the capacitor become charged with electrons. When the pulse of current stops the capacitor sends its stored electrons through the meter to keep the needle steady at one position until the next pulse. In electronics, this is called a "filter" because it smoothes out the pulses into steady current through the meter.

As an added feature, you can also measure the battery voltage. Just put the battery in the clip, switch your meter to volts, and plug the adapter into your meter. Now when you touch the pointed test probes together the meter will read the battery voltage.

PARTS LIST
(Catalog numbers here refer to 1977 Radio Shack Inc. catalog number 279;
a wide range of substitutions is possible.)

    1 non-insulated battery clip no. 270 379
    1 bakelite box no. 270 231
    1 Electrolytic Capacitor; 470 microfarads, 16 volts, no. 272 1007
    1 test leads with tip plugs to match your meter connections

QUARTZ OSCILLATOR TESTERS

By Wes Door, CMW

Before you turn this page thinking that this chapter is not for you because you are not really into quartz watch repairs, let me assure you this is (or can be) for you.

Although most watchmakers are involved in fitting new cells in quartz watches, some are also involved in other repairs such as installing new oscillators. Regardless of the extent of our repairs, one very useful piece of equipment is an Oscillator Tester. This may be purchased new or one may be easily made, as we will see shortly. By using this tester we can check a new quartz crystal to see what potential rate it will make after placing it in the watch. Since we can make this test in advance, it eliminates much trial and error testing.

Knowing that oscillating crystals vary in hertz (vibrations) is like hairsprings being made of different strengths for balance wheel movements. The shorter (or stronger) the hairspring, the faster the watch will run, all else being equal; the longer the hairspring (or weaker), the slower the watch will run. In these "tick-tock" watches we can vibrate a hairspring by suspending the hairspring (with balance attached), and allowing the lower balance staff pivot to barely touch a convex glass surface. We can actually count the number of balance wheel oscillations, especially in the older ones, with an 18,000 train (18,000 vibrations per hour equals 5 times per second) or 5 hertz.

However, in our quartz watches with oscillator quartz crystals vibrating over 32,000 hertz, it is impossible to count these by using this method. Besides, this extremely small oscillator piece of quartz crystal is enclosed in a small canister and all we can see are the two wires that extend from the canister. To replace an oscillating quartz crystal, we solder these wires to the appropriate place in the module, and our job is complete. Sounds simple . . . and it does not matter which wire goes to which end.

We do, however, have a few things to remember: We must select an oscillating crystal which physically will fit inside the space allotted (that is, either the same size or a smaller size than the original). This would be like saying putting any hairspring in as long as it is the same size or even smaller is OK.

It is this simple, but we should choose one whose lead wires match the original, as shown in Figure 1A. This illustration shows one wire on each end, while Figure 1B shows *both* wires are on the same end.

Although in many cases these lead wires are long enough to use either type, it would be a nicer looking job to use the one that properly matches the original. Since these replacement quartz crystals are readily available (generally between $1.00 to $2.00 each) it would be desirable to invest in a few. Maybe one or two dozen of each would be a nice start. Hopefully the oscillation values (hertz) of these will vary, which will allow us to select the proper one from this assortment.

Now the question comes: Why would we want these to vary in oscillations? After all, most quartz watches have a method of "trimming" (regulating). We simply turn a trimmer screw

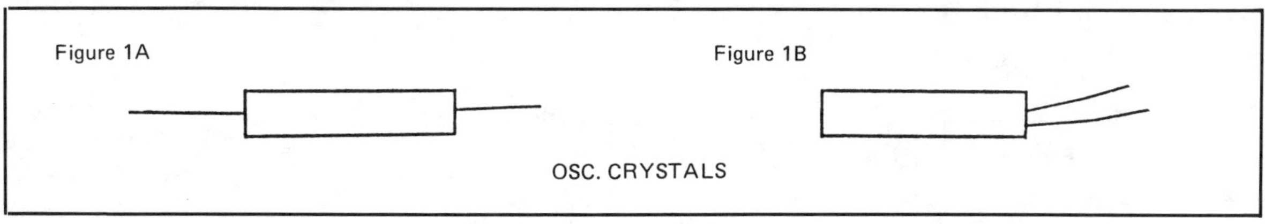

(in most modules) which is like pushing the regulator toward the fast or slow in "tick-tock" watches.

Let's suppose we receive a quartz watch that is off time and has no trimmer. In this case we must select a new oscillator crystal to compensate for this. If the watch runs 3 seconds fast per 24 hours we need a minus 3 second new oscillator crystal. Also, if a watch has a trimmer but cannot be brought into regulation, a new oscillating crystal will correct this error.

So we see we need an assortment of these oscillating quartz crystals. Since they are not pre-marked for us, we must be able to test these crystals before placing them into the watch. To accomplish this we need an Oscillator Tester which can be purchased or we can make one.

Here we are going to make one, but first I must give credit to Lou Zanoni of Zantech who first taught me how to do this. I know he sells these testers, maybe other suppliers do also, so I am not trying to give him a commercial. But we should be aware of unselfish people who help us in this new facet of our industry.

Now, to make an oscillator crystal tester we simply disconnect the oscillator quartz crystal and devise a means of re-connecting other crystals to be tested.

Step One: First select an old (or new) quartz watch (I prefer an LCD type watch, fairly small module).

Step Two: Be sure it runs and trim it if you have a quartz timer, and if not check it a few days, or borrow a friendly competitor's timer for about 5 minutes and trim it. *NOTE: After trimming, do not ever turn the trimmer again in this test model.*

Step Three: Remove the oscillator quartz crystal by cutting the lead wires close to the quartz crystal, thereby leaving the wires as long as possible.

Step Four: Add short pieces of wire and connect one alligator clip to each end.

Instead of alligator clips, one may make a more elaborate method for holding crystals to test, as shown in Figure 2. By enclosing the module in a clear plastic box and by presetting this module in the "seconds" mode it is easy to observe that the watch is running. This not only proves that the new crystal (which we are testing) is good, but it also shows that the power cell in our tester is OK. Now that we have a new crystal in our tester, let's let it run long enough to establish that its rate is correct, and then we may place it in the watch.

Other than a cell replacement, probably one of the easiest repairs we can make is the replacement of an oscillator quartz crystal. Of course, if we have a quartz timer, we would place our tester (with the new crystal attached) onto this quartz timer and note the rate, thereby preselecting the proper rate crystal for each job. If we are serious about quartz repairs, then Timers, Analyzers, etc. are a must. However, the quartz tester by itself can also be a useful tool.

OSCILLATOR CRYSTAL TESTER

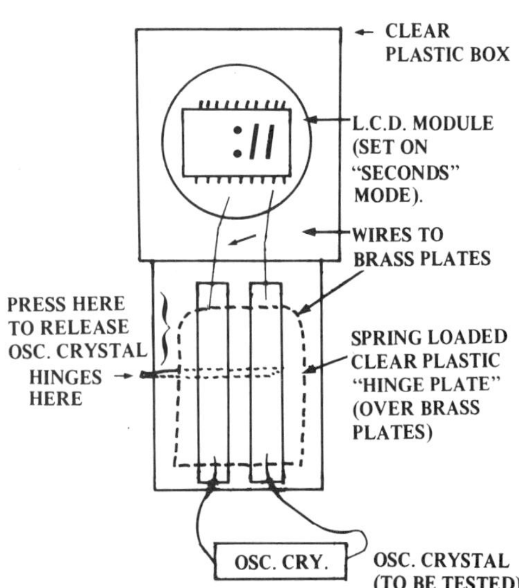

Figure 2

FEATURING THE MECHANICS
OF QUARTZ WATCHES

By James H. Broughton, CEWS

Much has been written about how to diagnose problems with quartz watches, most dwelling on the the electronic portion of the watch and little on the mechanical segment of quartz analog watches. In this chapter I will discuss some of the mechanical problems.

The calendar portion of a quartz analog timepiece should be no more of a problem to the watchmaker than it is on a mechanical watch. In fact, some of the calendars are identical to those on mechanical watches and the same parts do interchange.

With the calendar posing no particular problem we will concentrate on the time train. The time trains in quartz watches are very simple. They usually have three wheels and a rotor.

In their booklet, *Electronic Watches With Quartz-Crystal Resonators*, Ebauches SA depicts the two basic train types used in their quartz analog watches as shown in the illustration. Figure 1 shows the watch which has a sweep second hand as well as a calendar. Figure 2 shows a watch without sweep second drive or calendar. In all brands of watches these trains do require special care in handling.

Cleaning and a free train have always been important factors to the mechanical watch. Ultrasonic cleaning methods aided the watchmaker in achieving optimum results. While it is well known that cleanliness is a most important factor for mechanical watches, the quartz analog watch is many times more sensitive to dirt, lint, or obstructions in the train. There is no power on the train; therefore, the slightest piece of foreign matter in the train can stop the watch.

While most watchmakers understand the advantages of a clean watch, many forget that in order to get a watch clean you first must have clean solutions. I don't care what method of cleaning you use, even if it is stringing the parts on a wire, clean solution is a *must*. One simple procedure that may help in keeping the solution cleaner in the can is that when opening the can to fill a jar, close it again as soon as you are done. At the same time place a cover over the jar. This prevents airborne lint from getting into the solutions. Also, be sure to change solutions frequently.

While some watchmakers may disagree, I firmly believe that it is a must to completely disassemble the quartz analog watch for cleaning. I cannot find a reliable method of removing the metal particles from the stepper motor while it is completely assembled. The rotor sits in a hole of the stator and is very difficult to get to in order to remove the metal particles it attracts. Probably the only time I would want to put a stepper motor into a cleaning jar would be when I wanted to clean the jar out. It would collect the metal particles from the solution.

When removing the train bridge on a quartz analog watch, you should look at the wheels before removing them; it will make reassembly easier. Notice whether the pinions are

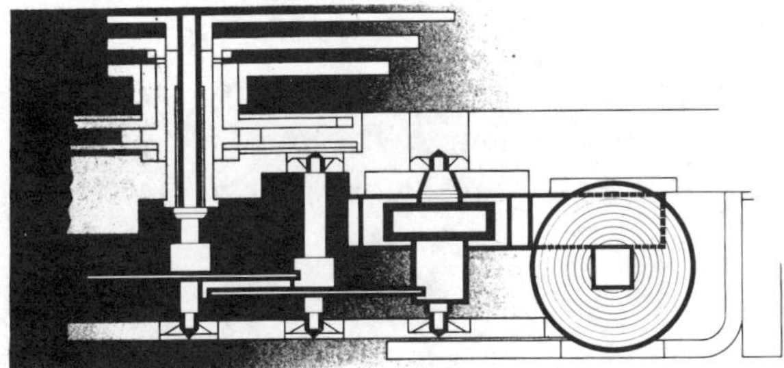

Figure 1

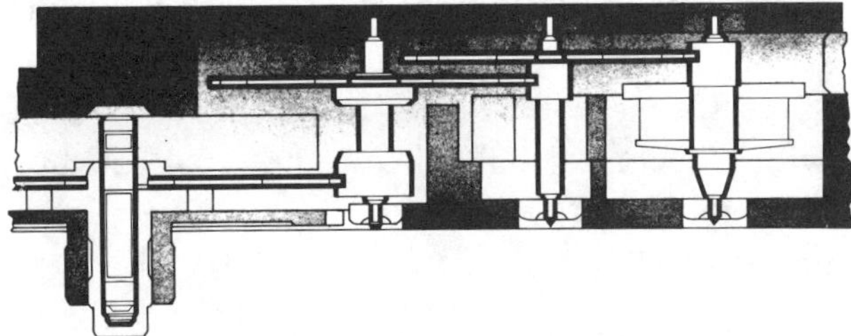

Figure 2

down or up. Notice the hack feature. Now, with the position of the train wheels in your mind, proceed with your disassembly; check each wheel as you remove it. When removing the rotor you should use nonmagnetic tweezers and after inspecting it, place the rotor in Rodico® One Touch or pithwood for safekeeping and/or protection.

The train bridge should be examined carefully, both before and after cleaning. Checking the bridge before cleaning is important if the watch was brought to you because it has stopped. Look close at the train bridge pivot holes for burrs. Burrs could be caused from a previous repair, but they will spell trouble for smooth operation of the train. Check for cracked jewels which could also be caused from a previous repair. Burrs are frequently caused by putting pressure on the train bridge when aligning the pivots up to the pivot hole. When aligning the pivots, it is very easy to put a slight scratch on the underneath side of the train bridge which can produce a small burr in the pivot hole. A burr may or may not cause a problem right away, but rest assured it will cause a problem sooner or later. Burrs can be removed with a small pivot broach, but care must be taken when removing them so that the hole is not enlarged.

If the train bridge is fully jeweled, check for a cracked jewel just as you would in a mechanical watch. Because the weight of the wheels is so light, about the only way the jewels will crack is when a repair person does not have the pivots properly seated before tightening the train bridge down. The replacement of a cracked jewel is the same procedure as it is with any other train bridge jewels.

When checking the train for freedom, you will not be able to spin the train of the quartz analog the same as you can the mechanical watch. The reason for this is simple. The rotor is being attracted to the stator by magnetic attraction. Just check the side-shake and the end-shake of the gears. You should use a pair of nonmagnetic tweezers to push the step motor around and let the magnetic attraction pull it the rest of the way. Doing this three or four times will permit you to see the train wheels advancing.

When assembling the quartz analog watch that has a sweep hand, be sure you pay attention to the hack feature after assembly to be sure it is working correctly. Special attention should be given these parts when disassembled so that you have a good idea how they go back together.

Quartz Analog Watches With a Sweep Second Hand

Special care should be taken when fitting a sweep second hand. Pull the stem out to the hand setting position. Place the sweep hand in position so the hand points exactly to 12 before installing it. By doing this, the sweep hand then will line up with each minute mark when the stem is pushed into the running position. It is very annoying to a customer to see the sweep hand stopping in between each minute mark. The customer will usually bring the watch back for correction.

When installing a new sweep second hand on a quartz analog watch, be sure you replace it with a balanced sweep hand. By this I mean one that has a tail on the back. Since there is no power driving the train, this kind of watch could stop when in a pendant position. The weight of the second hand, without a tail, is too much for some watches.

These tips about servicing the train portion of a quartz analog watch will help you be successful as you attempt to repair quartz analog watches. Keep in mind, AWI can supply you with a technical bulletin for the repair of most any watch that may cause you trouble. Don't hesitate to contact them.

ELECTRONIC DIAGNOSTIC PROCEDURE
ESA 9180-81-82-83

By Gerald G. Jaeger, CMW, CEWS, FAWI

When estimating or troubleshooting in any watch, whether it be mechanical or electronic, speed and accuracy are a prime concern. In the case of electronic, balance wheel, or stepping motor watches, it is extremely important because of the high cost of replacement of any of the electronic components. We must be able to recognize component failure with a few quick electronic checks. By the same reasoning, we must be able to quickly establish if our problem is electronic or mechanical.

In this chapter I will confine my specific procedures to the ESA 9180 series; however, the watchmaker who includes electronic and electric timepieces in his daily routine of repair will see that these steps can be modified and applied to most of these types of timepieces. In future articles, I will tie some of these similarities together and come up with a procedure applicable to most. The watchmaker attempting to repair this timepiece must have the Ebauches SA Technical Communication No. 27. This bulletin is available from either The Watchmakers of Switzerland Information Center (New York), or AWI Central (Cincinnati, Ohio).

The first consideration is: do we have sufficient power to run the watch?

Check No. 1 (Figures 1 and 2). The point here is that we remove the caseback only and remove neither the movement nor the cell from the watch. Set the VOM selector to read DC voltage on any range over 1.5V and at or under 10V. In the case of the meter in Figure 1, we would select the DC Volts range and set our selector at 3V (Check No. 1, Figure 1). Observing the proper polarity, we will place our + (red) meter probe on the positive (+) bridle No. 4401, which is indicated by a + on Figure 2. We will place our − (black) probe on the negative (−) bridle No. 4405, indicated by a − on Figure 2. Some of the earlier watches will use a mercury cell whose voltage when new and in good condition is 1.35V. The recommended cell is now a silver-oxide cell with a voltage of 1.55. This presents the possibility of two different voltage

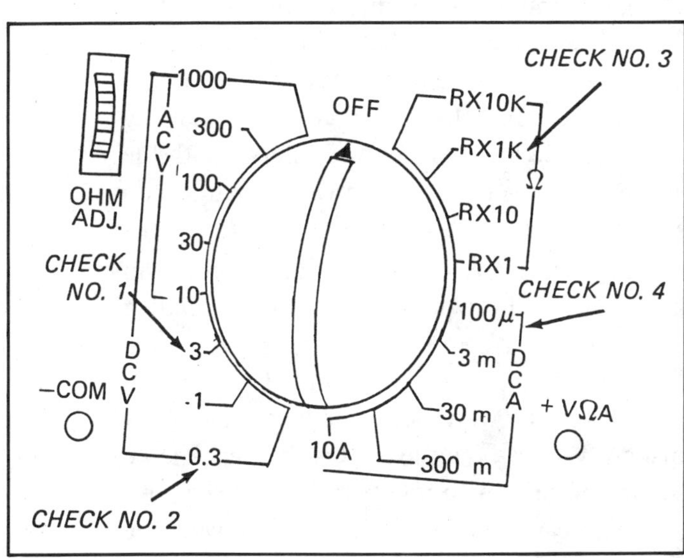

Figure 1

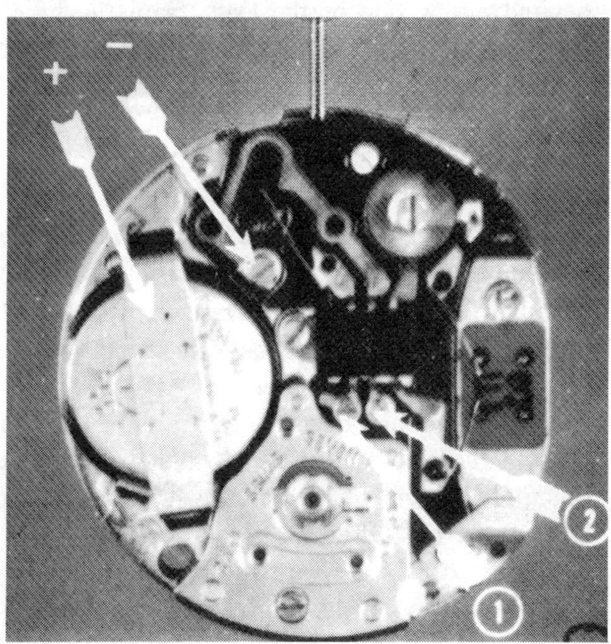

Figure 2

readings, either of which would be correct if the cell were still good. In the 1.35V cell, if the reading is below 1.10V and in the 1.55V cell if the reading is below 1.30V, the cell should be replaced.

Check No. 2. Having determined that the cell is good, or having replaced it with a new cell in good condition, we will quickly check the complete electronic unit. We can accomplish this by turning our range selector to the DC Volts range and choosing the lowest reading on our meter. On the meter pictured in Figure 1, we would set our selector at 0.3V (Check No. 2, Figure 1). Place one meter probe on the lead from the integrated circuit to the coil at the point indicated by No. 1 on Figure 2. Place the other meter probe at the point indicated by No. 2 on Figure 2. There is no need to observe polarity on this check, so either probe can be placed at either point. In Figure 2, we show the two screws removed, but in your watch, the screws would still be secured to the movement. This test is made with the crown pushed in at the running position. Our needle should alternately deflect to the right and to the left at one-second intervals. If it does, we know the complete electronic circuit is functional. If we do not get the proper needle deflection, we would have to complete the other electronic checks as indicated in Technical Communication No. 27. Assuming that after these checks we still have an inoperative electronic circuit, we would have to replace it. The point to remember is that we here are diagramming a few quick electronic checks for quick estimating and diagnosis.

Assuming we get the proper needle deflection in Check No. 2, we would continue with Check No. 3.

Check No. 3. This check is to quickly test the condition of the coil. It can be done with the cell either in or out. I find it easier to do with the cell in the watch. Remove the two connection screws which complete the connection from the integrated circuit to the coil. They are located at points No. 1 and No. 2 in Figure 2. Carefully spring the two leads up so that they do not make contact with the lead wires from the coil which are located directly under them. Choose the resistance range and set the selector of your meter at Rx1K. This is resistance times 1,000. Some meters may read Rx1000. Either is correct, as they are the same, K indicating 1,000. In this check it is not necessary to observe polarity, so our probes can be applied to the two leads directly under the leads from the integrated circuit. You can see that they extend beyond either side of the leads from the integrated circuit and can easily be probed without touching the leads from the integrated circuit. Our meter should read 3 on the ohms scale. This would tell us that our coil resistance is correct. The coil resistance can read from 2.5K to 3.2K and it will be correct. We can also test the resistance of the coils by removing the cell and applying our meter probes directly to the connecting screws No. 1 and No. 2 in Figure 3. Some may prefer this method, but the positive bridle (cell strap) may be a bit difficult to remove with the movement in the case. The meter reading should be the same as in the previously explained coil resistance test. If the coil is open or shorted, it must be replaced.

Check No. 4. This test will reveal if the watch is drawing or using more current than meets the manufacturer's specifications. It can be accomplished in two different ways, depending on the type meter you have.

The majority of the meters being used today do not have an independent power supply. On those that do, we can fit the recommended cell for the watch into the meter and use it to supply current to the watch when making a consumption test. On the meters that do not, we have to use the cell while it is in the watch as the source of current to power the watch and have it connected in series with the meter while doing do. This can be accomplished by removing the bridle No. 4401 (cell strap). This bridle acts as the connection between the + pole of the cell and the main plate which is the ground of the circuit. Set the selector of your meter at $100\mu A$ (100 microamps), Check No. 4 on Figure 1. With the cell in the movement place the positive lead of your meter on the back of the cell, which is the positive pole of the cell, and place the negative probe anywhere on the main plate; avoid touching any of the circuitry or plastic with the negative probe during this test. Your meter should read not more than 16 micro amps (16μA) when using a 1.35V cell and 19μA when using a 1.55V cell. This reading will be somewhat confusing, as the needle on your meter will be pulsing once each second, but you can read it at its peak. I have found the peak reading will read somewhat higher than the actual consumption. If you have a meter fitted with a condenser the needle will slowly move up to the actual consumption reading. Few watchmakers have a meter of this type, as it is an adaption which applies specifically to watchmaking and is relatively new.

If your meter has an independent power supply, merely insert the cell in the enclosure on the meter and select the lowest micro amp (μA) position and place your positive meter probe on the main plate, indicated by the + in Figure 3 and the negative meter probe on the negative bridle No. 4405 indicated by − in Figure 3. As indicated in Figure 3, the cell is not in the watch. Your consumption reading should be the same as the check using the cell while in the movement as the power supply.

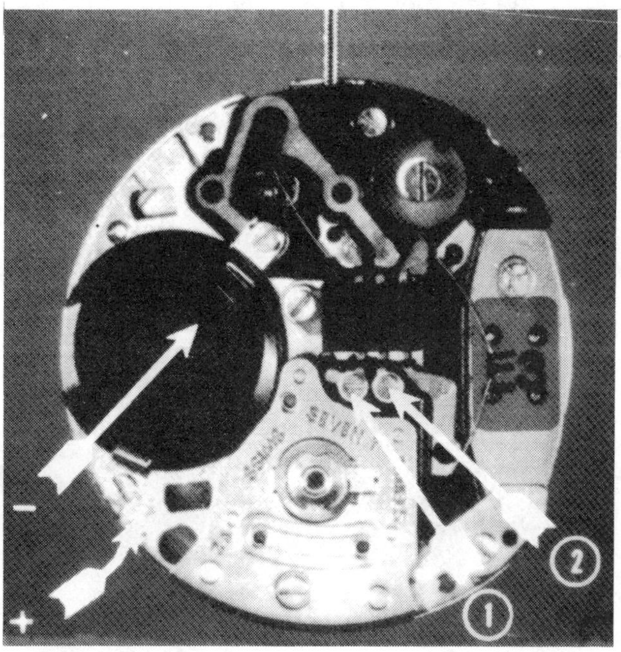

Figure 3

We have just explained how to make the consumption test in the ESA quartz resonator watch. In the ESA 9180 series we should have revealed a high current consumption prior to Check No. 4, as a train blockage will not be revealed by a consumption check, since there is not a mechanical connection between the electronic system and the mechanical system. The most common cause of a high consumption in this movement would be a low coil resistance.

Many manufacturers recommend a current consumption test. I personally doubt the value of this test, as in most cases we are merely confirming what we have already proved. In the case of the balance wheel in electronic watches, if a mechanical blockage or drag is present in the train, the damping test will reveal this. Excessive friction anywhere in the watch will also show up as a high current consumption, so both tests are proving the same thing. Frankly I feel the mechanical test is superior to the electronic test on any watch where there is a mechanical linkage between the balance and the train. In most cases, the resistance of the meters on the market today, when inserted into the watch circuit to measure consumption, will not provide enough current to bring the amplitude of the balance up to a high enough motion to read consumption. If the balance is not at its maximum swing, the consumption reading will be high. This test may be practical in the future when more watchmakers have meters specifically designed for this purpose. Until then I feel the damping test is superior. In fact, most balance wheel watches of electronic design and electrically powered cannot even be made to run when the present meters are connected into their circuit.

In the electronic watches where there is not a mechanical linkage between the resonator and the train, a high consumption reading will most always be the result of a lesser coil resistance than specified by the manufacturer.

Assuming that our electronic checks all work out, then we know our problem is mechanical. We are all professional watchmakers and are well trained at solving mechanical problems. The procedure we are trying to follow is very similar to that which should be followed in diagnosing problems in spring-driven watches. Checking the mainspring is equal to testing the cell. In both cases we have checked out our power supply. Checking the electronic circuit and the integrated circuit and coil is equal to checking the time train and escapement.

In all cases the checking must be made quickly and accurately in order for us to repair electronic watches profitably. We must determine if our problem is electronic or mechanical. Granted, these timepieces are new to many watchmakers, but with a little effort and dedication they too will become second nature.

The checks as I have outlined them are applicable directly to the ESA 9180 and 9181. A quick look at the picture of the component parts of the ESA 9182 and 9183 should enable you to make the same checks on that caliber.

REPAIRING AND REPLACING WATCHCASE PUSH BUTTONS
Part I

By Louis A. Zanoni

The watchcase push button is a spring-loaded plunger mounted on the side, front, or back of a watch case (see Figure 1). Its primary purpose is to make electrical contact with the watch module when pressed by the user. This is the means by which the watch is used and/or adjusted. When a watch is functioning properly, buttons are used to display additional information that is available within the integrated circuit, such as date, alarm time, or stopwatch function. Due to the high current drain of the LED display and the night light of an LCD, the buttons are also used to activate these features whenever desired.

The function of the button is to complete an electrical circuit to the IC when it is pressed. Generally, the IC is programmed to perform specific functions when the case button makes contact with the module. In some watches, the button moves two contacts together on the module to complete the circuit.

The IC is an electronic device, and it requires an electronic impulse in order to respond. The watches that have movable contacts mounted on the module do not require a metal button to complete the circuit. A plastic button can be used to move the two contacts together. This type is primarily used in the inexpensive plastic-cased watches.

The majority of watches use metal cases and metal buttons. It is, therefore, convenient to use the case to complete the electrical circuit from the button to the module. The plus terminal of the battery is usually connected to the case, and the metal button is press-fitted into the case, making electrical contact to it. Therefore, the plus battery terminal and the button are electrically connected together. When the button is pushed in, it makes contact with the switch terminal of the module. This connection does one of two things to the module: either it applies a plus voltage to the switch contact, which is electrically connected to the IC and the IC responds, or it shorts the negative voltage of the switch to the plus terminal of the case, cancelling out the negative voltage on the switch terminal. This causes the IC to respond.

The second method is the most common. It can easily be identified by the fact that the switch contacts have a negative voltage on them at all times.

Even though all circuits are not electrically the same, it is merely necessary to insure good electrical contact to the module for the module to respond, regardless of the method used.

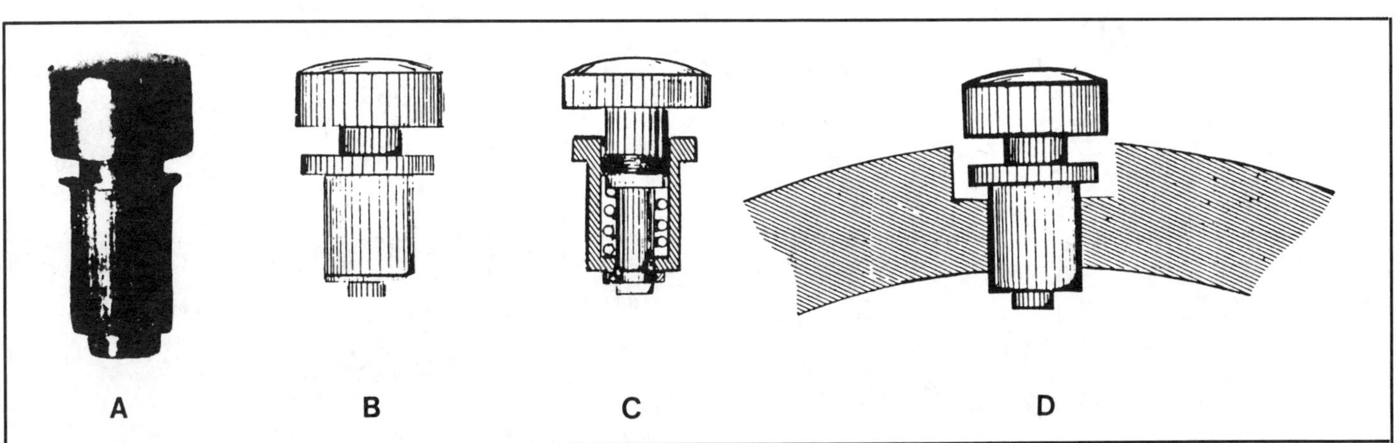

Figure 1

When troubleshooting a watch, it is important to determine that the problem exists with the button and not with the module before attempting to repair the button. Although most of the time, button problems are obvious—either broken, jammed, or missing—occasionally, the problems are electrical and associated with the module. For this reason, it is important to test the module before working on the buttons.

If the module is not functioning properly, repairing the button will not solve the problem.

Test Procedures

Regardless of the condition of the button, a systematic test procedure should be followed prior to doing any work on the button:

1. Physically examine all buttons and determine which ones require servicing.
2. Remove the caseback.
3. Microscopically examine all of the button contact points on the module before removing the module from the case.
   a. Are they contacting the module switch when pressed?
   b. Do they separate from the module switch terminals when released?
   c. Are the module contacts and button plungers clean?
   d. Are the module contacts properly connected to the module?
4. Remove the module from the case; remove the battery from the module and test the module completely (see "The Digital Watch Troubleshooting Guide," February and March 1981 issues of *Horological Times*).

Do not perform any work on the button until you are certain that the module is functioning properly. It would be wasted effort if the button were replaced and then the module were found to be beyond repair. The customer may not choose to pay the price of a new module and a button repair. **TIP: Test the module and examine it visually before repairing or replacing the button.**

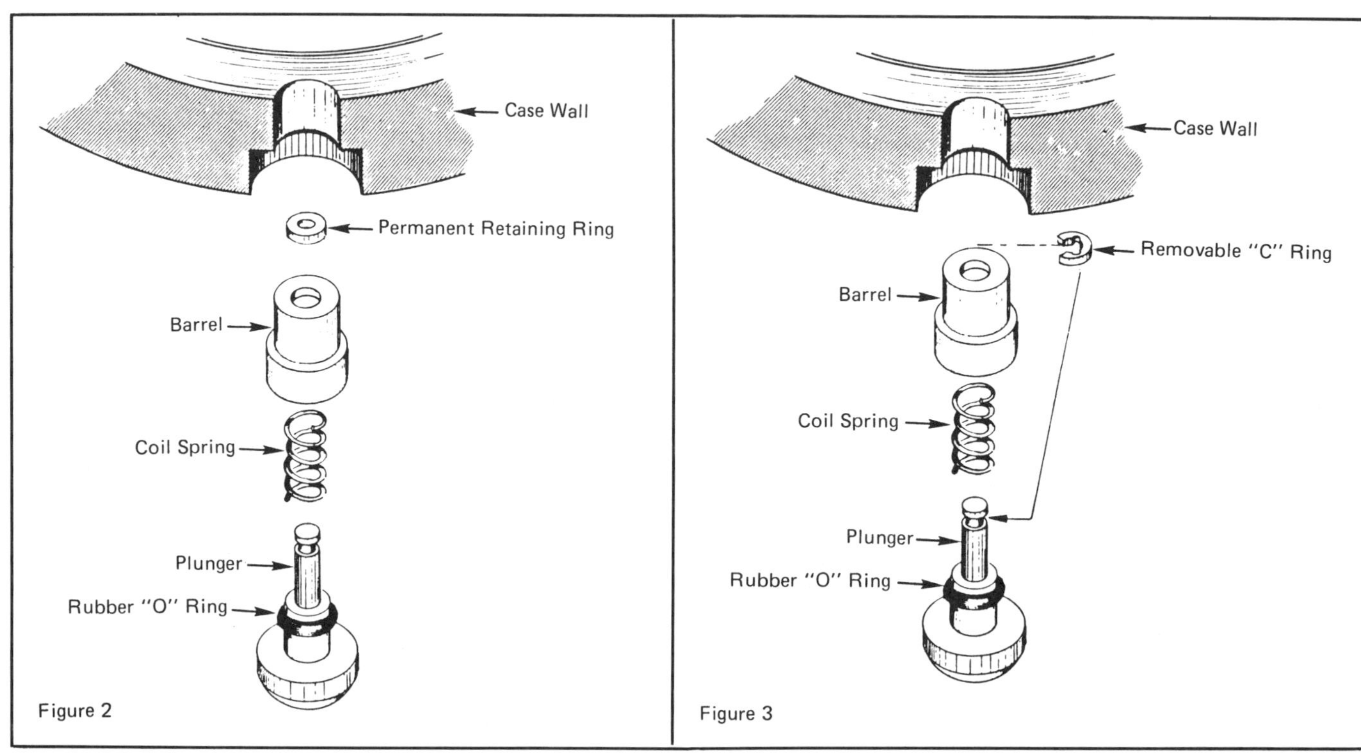

Button Structure

In order to service watch buttons, it is my opinion that a thorough understanding of them is necessary. Therefore, we will begin by describing the three basic types of button assemblies.

1. **The self-contained type.** See Figures 1 and 2. This type is a permanently assembled unit that contains all of the necessary components for a spring return movable plunger. See Figure 1, View C. This type is press-fitted into the wall of the case (Figure 1, View D).

2. **The C clip type.** See Figures 3 and 4. This type of button structure is very similar to the self-contained type, except for one major difference: it can be easily disassembled and reassembled for cleaning and lubrication. See Figure 3. Most of these buttons are press-fitted into the case wall in the same manner as the self-contained buttons.

   Other variations of this design use the case wall as the barrel of the button. It is very important to determine which type of button is being serviced.

3. **The springless type.** See Figure 5. This type uses a leaf spring on the side or front of the module to release the button. Its structure is the simplest of all. It consists of a solid metal or plastic plunger, pushed into a hole in the side or front of the case. A rubber "O" ring around the plunger prevents water from entering the module. A flange on the inside of the plunger prevents it from falling out. It is used to push on a leaf spring connected to the side or front of the module. The spring on the module returns the button to its normal position when released. It is *easily cleaned* and *easily lost*, so be alert when removing the module.

Failure Causes

The prime cause of button failure is dirt and contamination (see Figure 6A). The second is physical damage.

Many times there is nothing physically wrong with the button; it merely needs to be readjusted to make and break contact with the module under the appropriate pressure. Other associated button problems are caused by a buildup of corrosion on the contact point of the

Figure 4

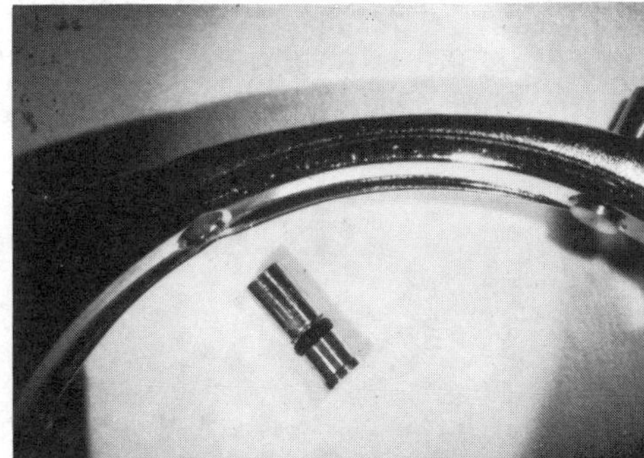

Figure 5

Figure 6A

Figure 6B

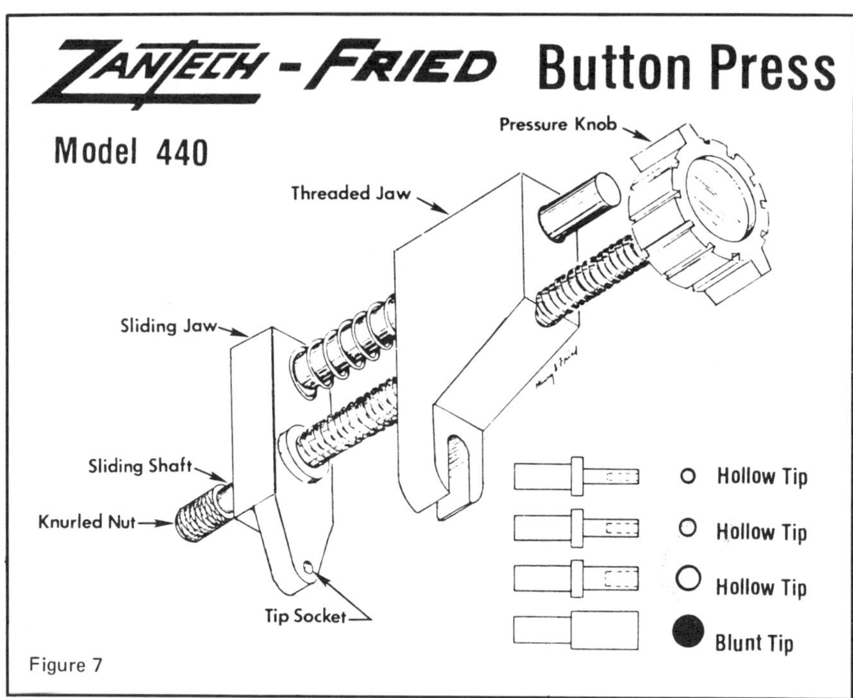

Figure 7

plunger or between the contacts on the module. All contact points must be thoroughly cleaned for proper operation. Many times the contamination is not visible to the naked eye. It is, therefore, recommended that one use a 10-power microscope for inspection. There are times when an even higher magnification is necessary. Cleaning contamination from metal contacts can easily be accomplished with a good, sharp screwdriver. Use it like a chisel or scraper. Any remaining residue can be removed with an ink eraser.

Removing the Press-fitted Button From the Case

The ability to remove and repair faulty buttons is advantageous because exact replacements are not always available.

Since most faulty buttons are caused by dirt and grime buildup in the moving portion of the button (see Figure 6A), it is important to remove the button from the case in order to clean it properly. See Figure 6B. Because exact replacement buttons are not always available, it is important not to damage the button when removing it.

Henry B. Fried and Louis A. Zanoni have designed and developed a unique button remover and adjustment tool known as the Zantech-Fried Button Press, Model 440 (see Figure 7). In the next chapter, we will describe the procedure required to successfully remove and insert buttons using this tool. The same basic principles could also apply to other types of button-removal tools.

REPAIRING AND REPLACING WATCHCASE PUSH BUTTONS
Part II

By Louis A. Zanoni

In the previous chapter, we determined the primary cause of push button failure to be dirt and contamination, with a secondary cause being physical damage. Since most button malfunctions are caused by dirt and grime buildup in the moving portion of the button, it is important to remove the button from the case in order to clean it properly. Because exact replacement buttons are not always available, it is important not to damage the button when removing it.

Henry B. Fried and Louis A. Zanoni have designed and developed a unique button remover and adjustment tool known as the Zantech-Fried Button Press, Model 440 (Figure 1). The following is the procedure required to successfully remove and insert buttons using this tool. These same basic principles could also apply to other types of button-removal tools.

Preparation

When the faulty button has been identified, examine the module side of the button very carefully with a 10X eye loupe or microscope. Be sure it is a press-fit type—most are (see Figure 2A). If the plunger (the moving part) of the button can be disassembled by removing a "C" ring from the end (Figure 3), it is not necessary to use a button press to remove it. This type can be removed by sliding the "C" off the end of the plunger and pulling out the plunger.

Tip

"C" rings are under spring tension when on the plunger. When removing them, be careful not to lose them. One method of preventing loss of these rings is to stretch a piece of thin, clear plastic over the case opening and push a tweezer or screwdriver through the clear plastic to remove the "C" ring. When the "C" ring slips off the plunger, it will stay in the case.

Selecting a Hollow Tip Driver

When it has been determined that the faulty button is a press-fit type, select a hollow driver tip that fits over the permanent retaining ring and the plunger (Figure 4). *Be sure the hollow end of the driver tip presses only on the barrel.* Do not press on the plunger of the button, the retaining ring, or the case wall. *Select a driver tip which fits the button.*

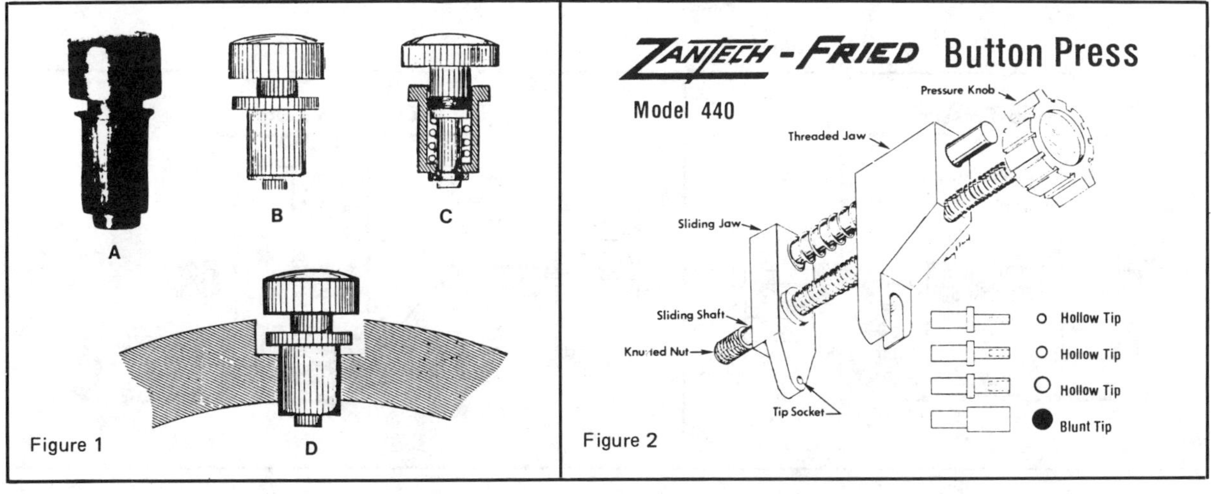

Symmetrical or Non-symmetrical Cases

Decide whether the pressure should be on the inside of the case (see Figure 5) or on the outside (see Figure 6). Use inside pressure on round cases when the buttons point toward the center of the case (Figure 5). Use outside pressure (Figure 6) for removing buttons which do not point toward the center, or when removing buttons from irregularly shaped cases.

Tool Adjustment

For inside pressure, screw the jaws of the press together so that the spring tension of the sliding jaw will hold the tool in position against the inside case wall. The hollow tip driver should be on the barrel of the button while the threaded jaw is against the opposite case wall.

Driver Tip Position

Place the appropriate driver tip in the socket of the sliding jaw (see Figure 7 for inside pressure or Figure 8 for outside pressure).

Button Removal–Symmetrical Cases

When the driver tip is properly in place and the threaded jaw is firmly against the opposite wall of the case (Figure 5), rotate the knurled nut to apply a moderate amount of pressure to the button barrel. *Before applying final pressure to the button, inspect the driver tip under a microscope or jewelers loupe to be sure it is making contact only with the barrel and not with the plunger or the case wall* (see Figure 4). When the tip is in its proper position, apply final pressure by rotating the pressure knob until the button falls out. To avoid damage to the driver tip, retract the threaded jaw before removing the tool.

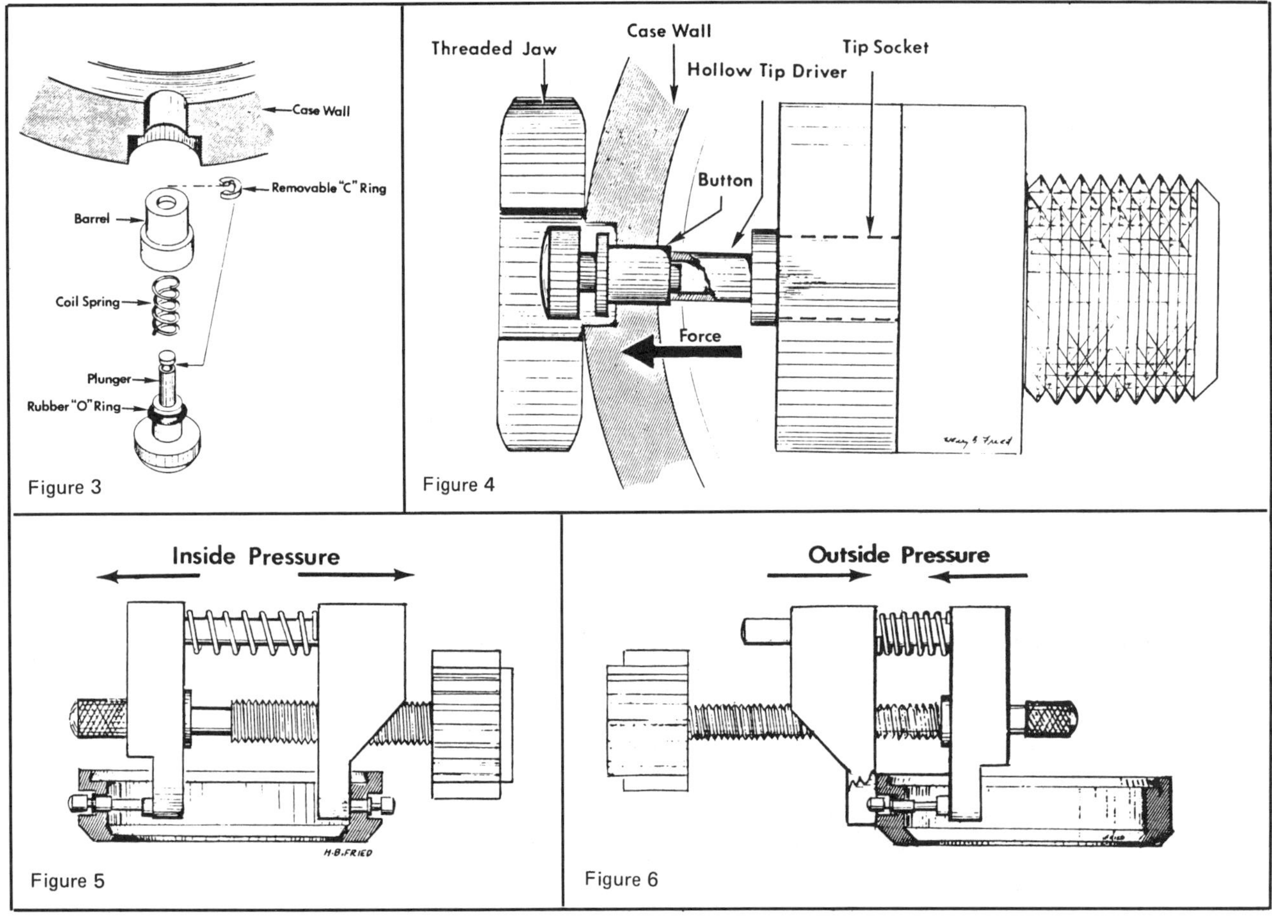

Button Removal—Non-symmetrical Cases

Place the appropriate hollow tip driver in the socket of the sliding jaw (see Figure 8). Adjust the jaw opening to the approximate size (Figure 6). Slide the hollow tip driver over the end of the button and screw the jaws together so that the threaded jaw is pressing on the outside of the case while the hollow tip driver is pressing on the barrel of the button (Figure 4). Screw the jaws together until the button falls out.

Cleaning the Button

In order to properly clean a contaminated button, it must be removed from the case. Once the contaminated button has been separated from the case, a variety of methods can be used to clean it. The method we found most economical and efficient is a diluted solution of ammoniated detergent in an ultrasonic tank for five to ten minutes. Although we have found it necessary at times to use a solvent cleaner similar to the L&R clock cleaner, the Zantech case and bracelet cleaner (Micro Clean) is a very effective cleaning agent for most types of contaminated buttons.

Ultrasonic vibrations are necessary to dislodge the contamination from within the spring mechanism. Soaking or jet steam cannot dislodge the contamination under the head as well as can good ultrasonic vibrations. It is necessary on occasion to physically dislodge large amounts of contamination by pressing and releasing the button a number of times between ultrasonic cleanings.

A thorough rinsing is also necessary to remove any traces of the detergent. Be sure to press and release the button a number of times to remove the detergent from the inner portion of the button. When it has been completely rinsed, it must be thoroughly dried. A paper towel works fine, and it can be used to wick out any remaining water within the mechanism. *If all of the water is not removed, the spring inside the button may rust.*

NOTE: Be sure to clean the case and especially the button cavity of the case before re-installing the cleaned button.

Replacing Buttons

After the original button has been thoroughly cleaned or when replacing a new button, merely locate it in the case by hand and gently press it into position. When it is properly positioned, press it firmly into the case wall by pressing it against the edge of a table or workbench. Normally, two-handed pressure is enough to firmly seat the button. When it is necessary to press the button deeper into the case, the button press should be used along with the blunt tip driver. *Do not apply pressure with the button press until the button is properly seated* (see Figure 9).

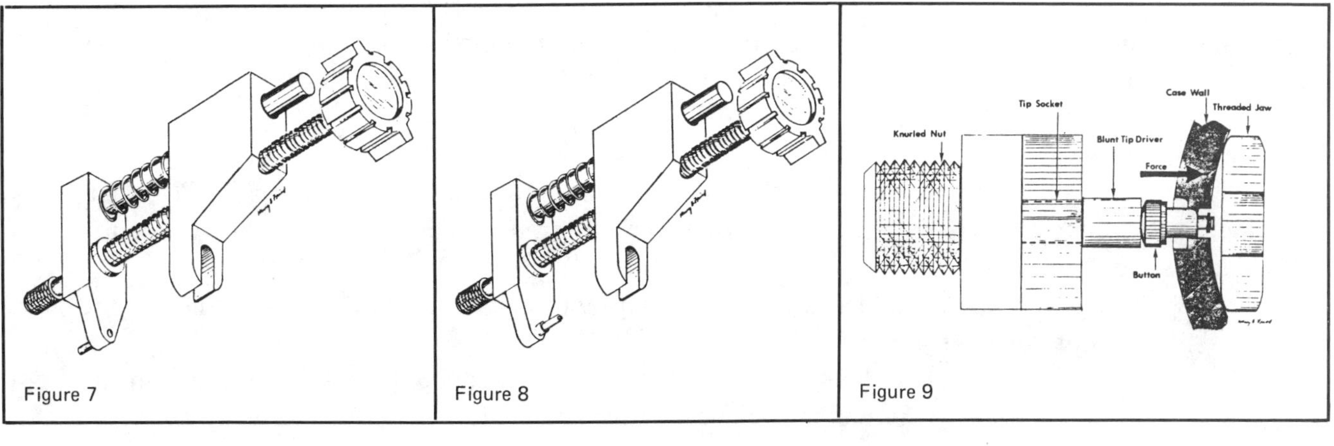

Figure 7

Figure 8

Figure 9

Button Depth Adjustment

The barrels of most press-fit buttons are slightly tapered so that they can be firmly fitted and positioned into the case wall. The button barrel diameter and the hole in the case should be precisely matched so that the button will be properly positioned when pressed into the case.

Button depth is very critical to the proper operation of the watch. If the button is in too far, it will make contact to the module continuously. If it is not in far enough, it will never make contact with the module. Therefore, it is occasionally necessary to make minor adjustments to the depth of a button. To push a button farther into the case, place the blunt tip driver on the inside of the tip socket (Figure 9). Position the blunt driver tip on the head of the button and press it to the desired depth by turning the knob. Each complete revolution of the knob moves the threaded jaw .9mm, or .035 inches.

CAUTION: Excess pressure on the button will damage the case, especially a thin-walled case.

When a button is pressed too far into a case, minor adjustments can be made by pushing the button partly out, using the same procedure required to remove a button. Keep in mind that one complete revolution of the knob moves the threaded jaw .9mm or .035 inches.

Lubricating Buttons

All types of buttons require some sort of lubricant because the cleaning process has a tendency to remove the original lubricant. Therefore, it is necessary to relubricate the buttons before placing them back into service.

Buttons of the **C clip** type can be easily lubricated with winding grease or silicon grease. The grease can be applied to the "O" ring prior to reassembling. The same can be done for the **springless** type of button.

The **self-contained** type of button requires special handling because it is a sealed unit. For proper lubrication, proceed as follows:

1. Fit the button into the case. Do not oil it until it is firmly and properly in place.
2. When the button is permanently in position in the case, extend the plunger into the case by pressing the button.
3. Place a small drop of heavy-duty clock oil on the part of the button that will be retracted into the case when you release the button. To evenly disperse the oil, press and release the button many times.
4. Remove any oil that may interfere with the electrical contacts of the switch.

Button Fitting Methods

When the original button is beyond repair and the exact button replacement is not available, it is occasionally necessary to modify the case, the module, or the button to achieve a proper fit.

Large Case Holes

When the hole in the case wall is too large, a number of solutions are possible. Depending on the difference in size between the case and the button, one or more of the following solutions may be applicable:

1. Raise burrs on the barrel of the button by rolling the button between a file and the workbench (see Figure 10).
2. Decrease the case hole size by deforming the inside wall of the case next to the button hole. One method would be to make center punch marks adjacent to the

Figure 10

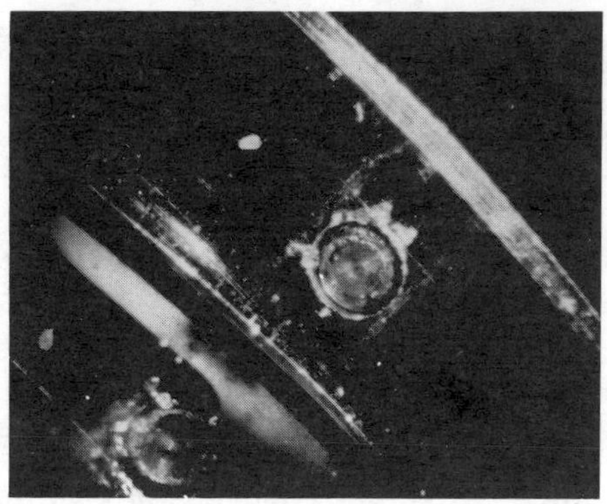

Figure 11

Figure 12

hole (Figure 11).

3. Use a small amount of Super Glue (or similar glue) on one side of the barrel of the button to cement it in place. Since the button must make electrical contact with the case, be careful not to insulate the entire button from the case with the cement (Figure 12).

4. Add tin-lead solder to the barrel of the button. Tin-lead solder will automatically conform to the shape of the hole when it is pressed into the hole. This is a favorite method. Contrary to common belief, the "O" ring is not damaged by soldering (see Figure 13).

5. A very large hole can be accommodated by epoxying the button in the hole with a thick paste, strong bonding epoxy. After the epoxy has firmly cured, electrical conductivity to the case can be achieved by painting a stripe of electrically conductive paint or epoxy from the edge of the barrel to the case wall (Figure 14).

NOTE: When deciding which approach to use, keep in mind that most buttons must make electrical contact to the case.

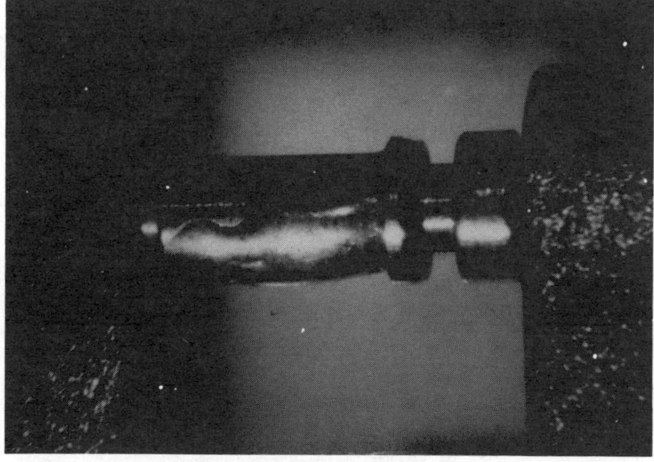

Figure 13

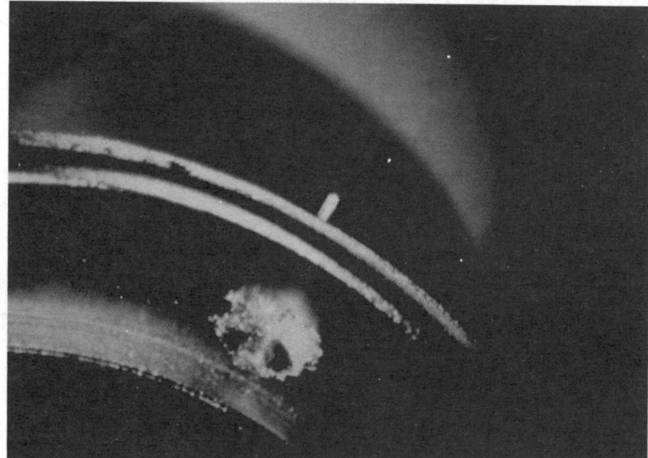

Figure 14

Small Case Holes

Whenever it is necessary to increase the hole size in the case to accommodate a new button, extreme care is necessary. The hole size must be very precise for a proper fit; therefore, the slightest increase in hole size will cause a mismatch.

The best method by which to achieve a good fit is to increase the hole size very slowly with a tapered reamer (see Figure 15). It is necessary to try fitting the button periodically during the reaming process until it is properly seated.

Two types of tapered reamers are available: the five-sided clock broach type (assorted sets are available from most clock material supply companies), and the high-speed steel type. The clock broaches are useful on brass cases. They are not suitable for cutting stainless steel cases. To enlarge a case hole in stainless steel it is necessary to use high-speed steel tapered reamers (see Figure 16). These are available from tool supply companies or your local watch material distributor.

Conclusion

Watches with buttons are here to stay, and as long as they are used by humans they will become damaged. It is our job to service watches. Therefore, the ability to service a button is an important part of the watch service business.

Once the technique has been learned, you will discover that it requires more time to read about how to replace a button than it does to replace one! This is profitable business—so don't let it get away.

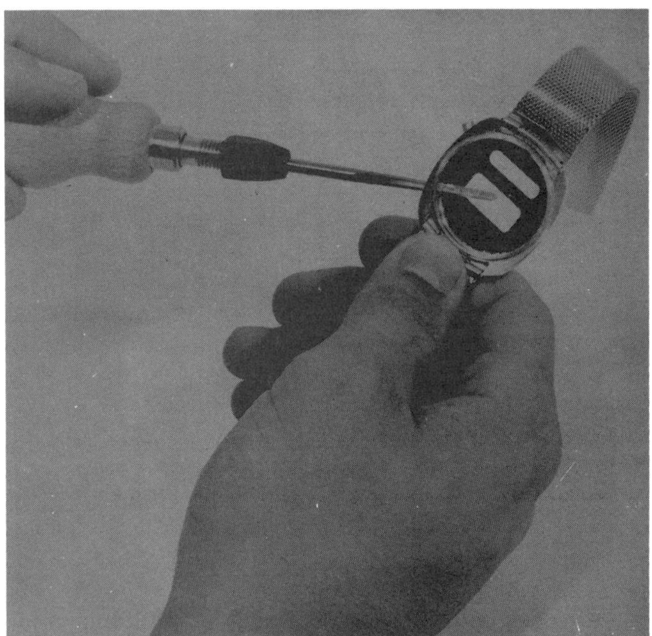

Figure 15

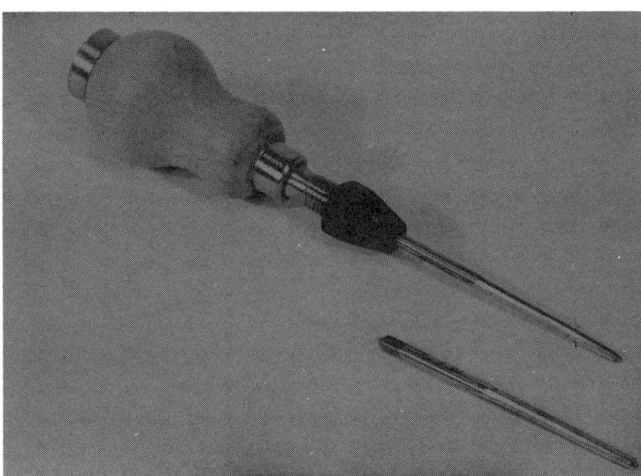

Figure 16

28 AU91 — GW. RH